Flatland

An Edition with Notes and Commentary

Flatland, Edwin Abbott Abbott's story of a two-dimensional universe, as told by one of its inhabitants who is introduced to the mysteries of three-dimensional space, has enjoyed an enduring popularity from the time of its publication in 1884. This fully annotated edition enables the modern-day reader to understand and appreciate the many "dimensions" of this classic satire with commentary on language and literary style, including numerous definitions of obscure words and an appendix on Abbott's life and work. Historical commentary, writings by Plato and Aristotle, and citations from Abbott's other writings work together to show how this tale relates to Abbott's views of society in late-Victorian England and classical Greece. Approaching the book from a mathematical stance, additional notes and illustrations enhance the usefulness of *Flatland* as an elementary introduction to higher-dimensional geometry.

William F. Lindgren is a professor of mathematics at Slippery Rock University. He is the coauthor of *Quasi-Uniform Spaces* (with Peter Fletcher).

Thomas F. Banchoff is a professor of mathematics at Brown University. He is the author of *Beyond the Third Dimension, Linear Algebra through Geometry* (with John Wermer), and an introduction to a new edition of Henry P. Manning's *The Fourth Dimension Simply Explained*.

Flatland

by Edwin A. Abbott

An Edition with Notes and Commentary by

William F. Lindgren
Slippery Rock University

Thomas F. Banchoff
Brown University

The Mathematical Association of America

CAMBRIDGE
UNIVERSITY PRESS

CAMBRIDGE UNIVERSITY PRESS
Cambridge, New York, Melbourne, Madrid, Cape Town, Singapore,
São Paulo, Delhi, Dubai, Tokyo

Cambridge University Press
32 Avenue of the Americas, New York, NY 10013-2473, USA

www.cambridge.org
Information on this title: www.cambridge.org/9780521759946

MATHEMATICAL ASSOCIATION OF AMERICA
1529 18th Street, NW, Washington, DC 20036-1358
www.maa.org

First published 2010

Printed in the United States of America

A catalog record for this publication is available from the British Library.

Library of Congress Cataloging in Publication data

Abbott, Edwin Abbott, 1838–1926.
Flatland : an edition with notes and commentary / Edwin A. Abbott, William F. Lindgren,
Thomas F. Banchoff.
    p. cm.
Includes bibliographical references and index.
ISBN 978-0-521-76988-4 (hardback) – ISBN 978-0-521-75994-6 (pbk.)
1. Fourth dimension.   I. Lindgren, William F.   II. Banchoff, Thomas.   III. Title.
QA699.A13   2009
530.11–dc22   2009025030

ISBN 978-0-521-76988-4 Hardback
ISBN 978-0-521-75994-6 Paperback

Contents

v

Acknowledgments

This book has its origin in the second author's thirty-one years of research into the life of Edwin Abbott Abbott. Persons who were interviewed for this project include relatives of Abbott's students and his closest friends, John Y. Paterson, Howard Candler, and William S. Aldis; Hilary Hammer, a solicitor at the legal firm Vizards at Lincoln's Inn Fields, who provided a copy of the wills of Abbott and his children; Peter Stanley Price, a nephew of Annie Fawcett, the sole heir to the estate of Mary Abbott; Katharina Wolpe and Lawrence Leonard, owners of Wellside, Abbott's house in Hampstead; at the City of London School: James Boyes, headmaster, Terry Heard, mathematics master, William Hallett, chief porter; David Singmaster, secretary of the London Mathematical Society; Sir Basil Blackwell, who was personally responsible for the republishing of *Flatland* in 1926; Frank V. Morley, the author of a review of *Flatland* in the *Saturday Review* (1926); and Hayward Cirker, John Grafton, and Banesh Hoffmann, who provided information about the republication of *Flatland* as a Dover paperback in 1952. Michael Holleran and Laura Dorfman, students at Brown University, were assistants during the early stages of this project.

Alexander K. Dewdney, Michele Emmer, Linda Dalrymple Henderson, Rosemary Jann, Barbara Mowat, Royal Rhodes, Joan Richards, and Rudy Rucker contributed talks to the *Flatland* Centennial Conference at Brown in 1984. Anthony Michaelis, the editor and publisher of the *International Science Review*, offered the opportunity to publish an extended article on Abbott, "From *Flatland* to Hypergraphics" (Banchoff 1990b).

The authors' collaboration began in the fall of 2000 when the first author was on a sabbatical leave at Brown. There we were ably assisted by the Mathematics Department manager, Doreen Pappas; secretaries Natalie Johnson, Audrey Aguiar, and Carol Oliveira; and the computer system administrator, Larry Larrivee. At Brown University Library, we enjoyed the services of Merrily Taylor, Samuel Streit, Patrick Yott, Robin Ness, and Ann Caldwell. During the summers of 2001 to 2005, we here helped in various ways by Brown students Steve Canon, Craig Desjardins, Ryan Roark, Ryan Roth, and Harry Siple. In July 2008, Michael Schwarz created the majority of the line drawings in the annotations.

At Slippery Rock University, the Mathematics Department secretary, Debra Dickey, was most helpful. Rita McClelland and Kathleen Manning at Bailey Library were consistently accommodating. Hans Fellner translated essays of Gustav Fechner. Richard Marchand created Figure 19.2 and gave advice on drawing others. Cleve Cooke proofread Steve Canon's rendering of the second edition of *Flatland* into LaTeX. Students whose comments and questions led to improvements in the annotations include Jason Lloyd, Valerie Long, Caleb Pardick, Amy Robinson, and Adam Wilcox.

We are deeply indebted to Don Albers, who encouraged us from the beginning; he and Jerry Alexanderson urged us to consider making our book a joint publication of the Mathematical Association of America and Cambridge University Press. It has been a great pleasure to work with our editor Lauren Cowles and our production controller Marielle Poss at Cambridge University Press, our production supervisor Shana Meyer at Aptara, and Paul Hightower, who copyedited the text.

We are especially grateful to our friend Terry Heard, the archivist at the City of London School, who provided hospitality during our several visits to London, answered countless questions, and offered corrections and amendments to several versions of the manuscript. We thank Richard Guy, who read the entire manuscript, found several errors, and made many helpful suggestions for improvements. We also thank an anonymous reader whose trenchant criticism led to a much leaner manuscript. Another anonymous reader made a number of useful suggestions concerning our treatment of *Flatland* and classical Greece.

Clareann Bunker and Kathleen Banchoff have been abiding sources of advice and encouragement. We received comments from several friends, including Peter Fletcher, Bernard Freydberg, Worthen Hunsaker, Ockle Johnson, Cindy Lacom, Peter Hanson, and Scott Taylor.

Joan Richards collaborated with the first author to produce the "mathematical biography of Edwin Abbott," found in Appendix B. Barbara Mowat identified the allusions to the works of Shakespeare in *Flatland*. Martin Gardner shared materials on the fourth dimension and *Flatland*. Karen Parshall offered her judgment on the review of *Flatland* in the *Oxford Magazine*. Lila Harper sent several contemporary reviews of *Flatland* and called our attention to H. G. Wells's letter to Priestley. Hester Lewellen shared her *Flatland* course materials. Jim Tattersall supplied information on Arthur Buchheim and sent Abbott's essay, "On the Teaching of the English Language." Peter Machamer introduced us to the work of Nicole Oresme. James Borg first alerted us to the existence of the Macmillan correspondence at the British Library. Tom Morley explained the physics of sound in *Flatland*.

We thank Jonathan Harrison at the library of St. John's College, Cambridge, which houses Abbott's papers; Margaret Harvey at City University of London, which keeps the Athenaeum Collection; and librarian David Rose and headmaster David Levin at the City of London School, as well as librarians at the British Library; Trinity College, King's College, and the University Library at Cambridge; the Yale University Beinecke Rare Book and Manuscript Library; Hollis and Houghton Libraries at Harvard University; the Folger Shakespeare Library; the University of Pittsburgh; the Providence Athenaeum; the University of Leeds; Bodleian Library, University of Oxford; and University College London.

We acknowledge permission from the following sources to reprint material in their control: The Master and Fellows of Trinity College, Cambridge, for the title page inscription to Howard Candler; The Harry Ransom Humanities Research Center, University of Texas at Austin, for Abbott's letter to Richard Garnett and H. G. Wells's letter to J. B. Priestley; the library of the University of Auckland for the quotation from the biography of William S. Aldis and the photograph of Aldis and Abbott; Brown University Library for the cover of the second edition of *Flatland* and the inscription to Aldis. The quotation from Vladimir Nabokov's *Lectures on Literature* is reprinted with the permission of Houghton Mifflin Harcourt, copyright © 1980. The obituary of Edwin Abbott is reprinted with the permission of the *Manchester Guardian*, copyright © 1926. Our failure to obtain a necessary permission for the use of any other copyrighted material included in this book is inadvertent, and we will correct it in future printings following notification in writing to the publisher of such omission.

It is altogether fitting that there be a "Cambridge *Flatland*." Abbott graduated from the University of Cambridge, as did most of his closest friends and many of his students. His son was a fellow of Jesus College, Cambridge, and his daughter attended Girton College. His last great work, the fourteen-volume "Diatessarica" series, was printed in Cambridge at the University Press. We believe that he would be pleased that Cambridge University Press is publishing *Flatland*, and it is to his memory that this edition is dedicated.

Introduction

Edwin Abbott Abbott's *Flatland* is the story of a two-dimensional universe as told by one of its inhabitants, a square who is introduced to the mysteries of three-dimensional space by a sphere. Since the time of its publication in 1884, Victorian customs have become obscure, the meanings of words have changed, and historical allusions, which were obvious to a contemporary, now require explanation. The present edition is intended to enable a modern-day reader to understand and appreciate the "many dimensions" of this classic satire. The extensive annotations to the text include mathematical notes and illustrations, which enhance the usefulness of *Flatland* as an elementary introduction to higher-dimensional geometry; historical notes, which show connections to late-Victorian England and to classical Greece; citations from Abbott's other writings and the works of Plato and Aristotle, which serve to interpret the text; commentary on close parallels between *Flatland* and Plato's "parable of the cave"; notes on the language and literary style of the book, including definitions of obscure words; and an appendix, which gives a comprehensive account of Abbott's life and work.

A Romance of many dimensions.

The word "romance" in *Flatland*'s subtitle means a prose narrative that treats imaginary characters involved in events quite different in time and place from those of ordinary life. The two-dimensional world of *Flatland*, inhabited by geometric figures, is manifestly different from that of ordinary life; nonetheless, a great majority of its early readers were well acquainted with the "physical space" of Flatland. Flatland's space is the familiar Euclidean plane, the principal object of study in Euclid's *Elements*, a fixture in the curriculum of Victorian public schools.

The phrase "of many dimensions" is a play on various meanings of "dimension." In the literal (geometric) sense of the word, *Flatland* is a primer on the geometry of higher dimensions. But Abbott was not a mathematician and did not intend to write a geometry text; he would be surprised to learn that his "romance of many dimensions" has become a standard introduction to higher-dimensional geometry. He meant "dimension" primarily in a figurative sense and, in this sense, *Flatland* indeed has many dimensions: It is an extended metaphor expressed in the language of mathematics; it is a satirical commentary on Victorian society; it is a geometric version of Plato's parable of the cave; it is an expression of religious principle; and it is an illustration of Abbott's theory that imagination is the basis of all knowledge. Another dimension of *Flatland*, Abbott's use of play, has frequently been misjudged. The book certainly abounds with cleverness and play of wit, and Abbott certainly intended it to be amusing. But those commentators who have characterized *Flatland* as a mere humorous trifle have badly underestimated a book that is written in the same spirit of playful seriousness that pervades the Platonic dialogues.

Flatland *as Victorian England*/Flatland *as classical Greece.*

Although Abbott was not the first person to posit a two-dimensional universe inhabited by geometric figures, he was the first to imagine such a space endowed with a highly developed social and political structure. Abbott's primary model for this structure is not that of late-Victorian England, which is unquestionably the target of his satire, but rather that of classical Greece. Abbott's contemporaries would have found Flatland's society as familiar as its space, for the traditional public school education placed a heavy emphasis on the Greek and Roman classics.

There was a widespread conviction among Victorian writers that the Greeks had been like the English and "that the historical situations of the two civilizations were essentially similar. Although this attitude did not survive much beyond the first quarter of the twentieth century, it was fundamental to Victorian intellectual life and determined the outlook of much Victorian scholarship, criticism, and commentary on the Greeks." To maintain the similarity between the two civilizations, writers like Matthew Arnold had to rationalize away fundamental differences and ignore morally distasteful elements in Greek society. Others (with whom Abbott would have agreed) argued that classical Greece had not been like England, and England should not model itself on classical Greece (Turner 1981, 11, 61, 252).

In writing *Flatland*, Abbott used "historical imagination" not to reconstruct the past but to reconstruct the present in the past. He devised an extended geometric metaphor by projecting late-Victorian England onto a two-dimensional space with a "civilization" in various ways similar to that of classical Greece. Further, he heightened his satirical commentary on the present by making prominent in this imaginary civilization some of the very aspects of classical Greece that its Victorian apologists had rationalized away – for example, slavery, a rigid class system, misogyny, and ancient forms of social Darwinism.

Flatland *and Plato's parable of the cave.*

Several writers have noted the most significant "Greek connection" in *Flatland* – its parallels with Plato's parable of the cave. In the seventh book of the *Republic*, Socrates describes a cave containing prisoners who have been kept fixed since childhood by bonds on their legs and necks. Their bonds prevent them from turning their heads, and so they cannot see the fire burning above and behind them. Between the fire and the prisoners there is a rampart bordered by a low wall on which certain men place various artifacts; the fire casts shadows of these objects on the cave wall in front of the prisoners. These shadows and the voices of the men on the rampart are the only "reality" that each prisoner knows (*Republic*, 514a–518b).

In *Flatland*, Abbott has amplified Plato's metaphor and rendered it into the language of geometry by substituting a two-dimensional plane populated by geometric figures for the cave and its prisoners. Although Abbott does not explicitly acknowledge the source of his model, and the details of the two stories are by no means identical, the derivation of *Flatland* from Plato's parable seems unmistakable. Most significant, each text contains a metaphorical account of both the nature of the human condition and the journey of an individual soul from ignorance to knowledge.

Abbott's "geometrization" of the cave parable is particularly appropriate because of the importance that Plato attaches to mathematics. Plato included geometry in the educational curriculum for the guardians of the State because it "is the knowledge of the eternally existent" and tends to draw the soul to truth, and is "productive of a philosophical attitude of mind, directing upward the faculties that are now wrongly turned earthward" (*Republic*, 527b).

Precursors of Flatland.

The study of higher-dimensional geometry began with works by Hermann Grassmann (1844), Arthur Cayley (1846), and Bernhard Riemann (1854); by the time *Flatland* appeared, hundreds of articles on the subject had been published. Interest in higher-dimensional spaces was by no means confined to the scientific community, and a number of these articles were directed at a general readership. Several writers of popular essays on higher dimensions illustrated the difficulty of understanding four-dimensional space by portraying two-dimensional beings living on a surface and unable to perceive anything of three-dimensional space. This dimensional analogy, which is fundamental to *Flatland*, was used by three writers whose works might have influenced Abbott: Gustav T. Fechner, Hermann von Helmholtz, and C. Howard Hinton.

The German psychologist and philosopher Fechner was the first person to use "flatland" as a device for understanding higher dimensions. In two essays, *"Der Schatten ist lebendig"* ("The shadow is alive") and *"Der Raum hat vier Dimensionen"* ("Space has four dimensions"), first published in 1845, he describes a shadow man capable of moving about on surfaces and interacting with other shadows. Abbott, who was fluent in German, might have read them in *Kleine Schriften*, a collection of Fechner's satirical essays (Fechner 1875, 243–276).

Between 1868 and 1879, the German scientist and philosopher Helmholtz published several lectures and papers with similar titles and contents that considered how humans come to understand the nature of space. He used an example of two-dimensional beings whose movements are confined to the surface of a solid to argue that our notion of space is not, as Immanuel Kant supposed, an *a priori* intuition but rather is determined by our experience. Abbott could have seen the essay, "The origin and meaning of geometrical axioms," which appeared in the English philosophical journal, *Mind* (1876).

Hinton has been described as a "hyperspace philosopher." His essay "What is the fourth dimension?" (1880, 1883) is the first of several that he wrote to popularize four-dimensional space. Abbott might have learned of Hinton's essay from his friend Howard Candler, who was the mathematics master at Uppingham School where Hinton was the science master between 1880 and 1886.

Another precursor of *Flatland*, an essay entitled "A new philosophy," appeared without attribution in the November 1877 issue of the *City of London School Magazine*. The thesis of this satirical essay is that mathematics is the only "science" that can provide an unshakable foundation for a philosophy or religion. According to the proposed "Geometrical Philosophy," the universe is composed of an ascending chain of spaces, each having its dimension one greater than that of its predecessor, as well as a descending chain of spaces, each having its dimension one less than that of its predecessor. Although the *City of London School Magazine* was a student publication, it is possible that Abbott himself was the author of this noteworthy essay (New Philosophy 1877; Valente 2004).

Abbott could have found the idea of using a mathematical setting for his story in Lewis Carroll's *Dynamics of a Parti-cle*. Carroll prefaces this small pamphlet of political satire with a brief account of a love affair between a pair of linear creatures moving across a plane surface. This preface, he says, illustrates "the advantage of introducing the human element into the hitherto barren region of mathematics" (Carroll 1874).

Edwin Abbott Abbott.

The author of *Flatland*, the eminent biblical and English scholar and Victorian headmaster Edwin Abbott Abbott, was born in 1838 in Marylebone, where his father was headmaster of the Philological School. Abbott was educated at the City of London School (CLS) under George F. W. Mortimer before entering St. John's College, Cambridge, where he was senior classic and senior Chancellor's Medallist in 1861. He was ordained deacon in 1862 and priest the following year. After teaching briefly at King Edward's School, Birmingham, and Clifton College, Bristol, he returned to CLS as headmaster in 1865. The school that he inherited from Mortimer was highly regarded, and under Abbott it became one of the best day schools in England. He reformed the traditional curriculum, brought new methods of instruction, and improved the quality of the assistant masters. Abbott was a gifted teacher who sent a large number of students to Oxford and Cambridge. At the same time, as an administrator he ensured that the greater number, those not meant for the universities, received a sound general education.

His student and biographer, Lewis Farnell, asserted that Abbott's "claim to be remembered must chiefly rest upon what can only be called his genius for teaching." Nonetheless, the enduring interest of his life was the problem of presenting Christianity to his contemporaries in a way that would ensure the permanence of traditional beliefs without requiring the acceptance of miracles (Obituary 1926b). He "retired" at the age of 51 and devoted himself to biblical scholarship; between 1900 and 1917, he published an immensely detailed, fourteen-volume study of the four Gospels.

Various commentators have remarked that *Flatland* seems out of place with the rest of Abbott's literary output; nevertheless, it is quite similar to two other pseudonymous, first-person accounts that Abbott wrote: *Philochristus* (1878), the story of a Pharisee in the early first century, and *Onesimus* (1882), the story of the Greek slave in St. Paul's Epistle to Philemon. In these stories, as in *Flatland*, a protagonist is transformed by the revelation of a being of a higher order, but when he attempts to spread this good news, he meets with frustration and even persecution.

The first and second editions of **Flatland.**

The first edition of *Flatland* probably appeared in late October 1884. *Flatland* is on Seeley and Company's "List of New Books" in *The Literary Churchman* (24 October 1884, 452). Abbott autographed copies to several friends in October 1884, and the earliest review appeared in *The Oxford Magazine* (5 November 1884) (see Appendix A4).

We do not know how many copies of the first edition of *Flatland* were sold. In the nineteenth century, publishers typically issued 500 to 1,000 copies, and the great majority of books never went into a second edition. In any case, the first edition sold out quickly, and a second edition was published. The title page of this edition is dated 1884, but the following evidence suggests that it did not appear until 1885: The significant changes to the text are based on a letter, which the Square sent to *The Athenaeum* in response to a review of *Flatland*, and the reviewer's response to that letter. Both the Square's letter and the reviewer's response appeared in *The Athenaeum* on 6 December 1884 (Appendix A2). An advertisement for Seeley and Co. in the *Times* of 21 January 1885, describes *Flatland* as "just published" with no mention that it is a second edition. Abbott inscribed a second edition to Howard Candler in February 1885. The first advertisement for *Flatland* in the *Times* that mentions a second edition appeared in the 18 March 1885 issue.

Subsequent editions.

Flatland had been long out of print in England when it was reissued in June 1926 by Sir Basil Blackwell. The text of Blackwell's edition was only slightly different from the second edition, and it had much the same appearance as the original. This edition included an introduction written by the physicist, William Garnett, who was among Abbott's first pupils at the City of London School. In his introduction, Garnett cites an essay written six years earlier in which he characterized Abbott as a prophet who had foreseen the relevance of the dimensional analogy for understanding the passage of time in relation to space.

In 1885, Roberts Brothers of Boston issued the first American edition of *Flatland*, essentially the uncorrected first edition with Americanized spelling. In 1898, Roberts Brothers was acquired by Little, Brown and Co., which continued publishing *Flatland* until the middle of the twentieth century. Much of the popularity of *Flatland* in the United States is attributable to a co-founder of Dover Publications, Hayward Cirker, who chose *Flatland* as one of his firm's first titles in mathematics in 1952. The publication of the Dover *Flatland* made the second edition readily available to American readers for the first time. In the past thirty years, publishers have issued dozens of "editions" of the book, which differ from one another only in their introductions. The first translation of *Flatland*, the Dutch *Platland*, appeared in 1886; since then, it has been translated into sixteen other languages.

This edition of Flatland.

The text of *Flatland* that follows is the second edition with one change: The Preface to that edition is the Epilogue to this one. We have made this change because this Preface/Epilogue can be properly understood only by a person who has read the rest of the text. Furthermore, it really is an epilogue – the concluding section of a work in which a character, at a somewhat later time, reflects on the preceding events and gives additional details, which serve to interpret the story. Finally, reading this section before reading the text would spoil the effect of the narrative in which the Square describes his passage from the unenlightened "Square he once was" through his "initiation into the mysteries of space" and his subsequent "miserable Fall."

In his essay "Good readers and good writers," Vladimir Nabokov maintains: "In reading, one should notice and fondle details ... Curiously enough, one cannot *read* a book: one can only reread it. A good reader, a major reader, an active and creative reader is a rereader" (Nabokov 1980, 3). It is to the rereaders that we address this annotated edition of *Flatland*.

"O day and night, but this is wondrous strange"

FLATLAND

A ROMANCE OF MANY DIMENSIONS

By A Square

No Dimensions
POINTLAND

One Dimension
LINELAND

Two Dimensions
FLATLAND

Three Dimensions
SPACELAND

LONDON
SEELEY & Co., ESSEX STREET, STRAND
Price Half-a-Crown

"And therefore as a stranger give it welcome"

Notes and Commentary

Cover. As he did with *Philochristus* and *Onesimus*, Abbott wrote *Flatland* in an archaic style to create an impression of antiquity. In the case of the first edition of *Flatland*, the impression begins with its cover, which is made of vellum (a fine kind of parchment prepared from the skins of calves) wrapped on cardboard. Abbott may have chosen this cover as a reminiscence of the principal form of the "book" in the ancient world – a roll made of glued sheets of papyrus wound about a wooden stick and often kept in a parchment cover.

Cover epigraph. The epigraph, which is taken from *Hamlet* 1.5, points toward and provides commentary on *Flatland*'s central event, the nocturnal "appearance" in Flatland of a sphere, who has come to initiate the Square into "the mysteries of space." The first line, "O day and night, but this is wondrous strange," is spoken by Horatio, who has just seen the ghost of Hamlet's slain father appear and disappear. In his reply, "And therefore as a stranger give it welcome," Hamlet alludes to the ancient practice of hospitality to strangers. Shakespeare is urging us not to be limited by Horatio's rationalism but, together with Hamlet, to welcome "things not dreamt of in (our) philosophy."

It was natural for Abbott to select epigraphs from the works of Shakespeare. He established his scholarly reputation with the publication of *A Shakespearian Grammar*, which he wrote to furnish students of Shakespeare with an account of the differences between Elizabethan and Victorian English.

A Square. Edwin Abbott Abbott's name contains the surname of both of his parents, Edwin Abbott and Jane Abbott, who were first cousins. The pseudonym Abbott has chosen is a pun – it refers to his own initials, EAA = EA², as well as the modest social status of the "author" who is an ordinary square. *Flatland*'s everyman narrator does not tell us the names of any of his contemporary Flatlanders; nor does he tell us his own name – he is "a square" ("A Square" as it is typeset on the title page), not A. Square.

Abbott made the Square the author of *Flatland* not to avoid responsibility for the book but rather to present it as the memoirs of a two-dimensional being. The anonymous publication of books was common in the nineteenth century, but books that received any degree of celebrity were typically attributed within a few months. The first public indication that Abbott was the author of *Flatland* appeared in the Literary Gossip column of *The Athenaeum* (see Appendix A2, footnote 3). Abbott himself may well have been the source of this item. Certainly, he was the one who revealed that he was the author of *Philochristus*: "I shall publish it anonymously: but shall carefully let it be known that I am the author, for there are reasons why (though I may not like to be abused by *name* in the religious papers) I have no right to shirk the odium of heterodoxy, for the book is heterodox" (Abbott 1874).

Seeley & Co. The firm that published *Flatland* was owned by Richmond Seeley, the second son of Robert Benton Seeley and an elder brother of Abbott's friend and mentor, John R. Seeley. In 1857, Richmond Seeley took control of his father's share in the family business, which had been founded about 1784. Seeley continued the

FLATLAND

A Romance of Many Dimensions

With Illustrations

by the Author, A SQUARE

" *Fie, fie, how franticly I square my talk!* '

LONDON

SEELEY & Co., 46, 47 & 48, ESSEX STREET, STRAND

(*Late of* 54 FLEET STREET)

1884

Cover notes continued

traditions of the firm, which under his father's influence had acquired a strong tone of evangelical churchmanship.

In addition to *Flatland*, Seeley published ten of Abbott's books – *Bacon and Essex: A Sketch of Bacon's Earlier Life* and nine school books, including *English Lessons for English People* (with John R. Seeley) and *How to Write Clearly*. Abbott published these books at his own risk. He employed Richard Clay & Sons to produce the printed pages. Then Seeley & Co. bound, distributed, and advertised the books for a percentage of the revenue from sales. The income from the sales of his school books enabled him to retire in 1889 at the age of 51 (Abbott 1877e).

Price Half-a-crown. A half crown was $2^{1}/_{2}$ shillings, or $1/_{8}$ of a pound. Although changes in relative values of goods make it impossible to translate this price into a modern value, a rough indication is given by Leone Levi's estimate that in 1884 the average daily income of an English working man was somewhat less than 3 shillings (Levi 1885, 2–4).

Title Page.

Title page epigraph. "Fie, fie, how franticly I square my talk" means "how madly I adjust my language." It is taken from Shakespeare's *Titus Andronicus* 3.2, where Titus responds angrily to his brother's urging him to moderate his language of grief and despair. This epigraph is a play on the verb "to square" and the "name" of *Flatland*'s narrator. It refers to the Square's struggle to make his narrative description consistent with the "reality of Flatland," and it alludes to Abbott's own efforts to write *Flatland* in "the language of the Square."

Abbott was a philologist in the literal sense, a lover of words, and he took great care in constructing the Square's language, which is not merely the ordinary English of the late nineteenth century. It includes archaisms from Elizabethan English, biblical diction, mathematical and geometrical vocabulary, and a number of words peculiar to the "idiom of Flatland." The prose occasionally borders on the poetic; alliteration and other rhetorical forms are common. There is a large element of wordplay, including several clever puns, which, like this epigraph, often call attention to a noteworthy aspect of the text.

To

The Inhabitants of SPACE IN GENERAL

And H. C. IN PARTICULAR

This Work is Dedicated

By a Humble Native of Flatland

In the Hope that

Even as he was Initiated into the Mysteries

Of THREE Dimensions

Having been previously conversant

With ONLY TWO

So the Citizens of that Celestial Region

May aspire yet higher and higher

To the Secrets of FOUR FIVE OR EVEN SIX Dimensions

Thereby contributing

To the Enlargement of THE IMAGINATION

And the possible Development

Of that most rare and excellent Gift of MODESTY

Among the Superior Races

Of SOLID HUMANITY

Dedication.

H. C. in particular. In *Apologia* (1907), Abbott explicitly identified his closest friend, Howard Candler, as "the 'H. C.' to whom *Flatland* was dedicated many years ago." When he inscribed the title page of a copy of *Flatland* for Candler, Abbott wrote,

To H. C. in particular
from the Square.
Oct. 1884

That volume was given in 1969 to the library of Trinity College, Cambridge University, by Christopher Candler, a grandson of Howard Candler.

Initiated into the mysteries. Abbott uses the initiation rituals of the ancient Greek mystery cults as a figure for the Square's passage from intellectual darkness into light and his subsequent inability to describe his experience to others.

Imagination. For Abbott, imagination is the basis of all knowledge. In *The Kernel and the Husk*, he maintains that our knowledge of the external world and ourselves comes not from sensations as interpreted by reason but, at least to a large extent, from sensations as interpreted by imagination.

Modesty. In *The Spirit on the Waters*, Abbott says that an illustration set in geometric space may lead us to wider views of possible circumstances and existences, and thereby "develop in us modesty, respect for facts, a deeper reverence for order and harmony, and a mind more open to new observations and to fresh inferences from old truths" (Abbott 1897, 32–33).

PART 1

THIS WORLD

"Be patient, for the world is broad and wide."

Part I: THIS WORLD

"Be patient, for the world is broad and wide." (*Romeo and Juliet* 3.3) These words are spoken by Friar Laurence in an effort to comfort Romeo, who has been banished from Verona, by assuring him that the world outside Verona is spacious. Romeo responds:

> "There is no world without Verona walls,
> But purgatory, torture, hell itself.
> Hence – banished is banish'd from the world."

Romeo's insistence that nothing exists outside the world of his experience is a theme that is repeated in the text by inhabitants of Pointland, Lineland, Flatland, and Spaceland.

§1
Of the Nature of Flatland

I CALL OUR WORLD FLATLAND, not because we call it so, but to make its nature clearer to you, my happy readers, who are privileged to live in Space.

5 Imagine a vast sheet of paper on which straight Lines, Triangles, Squares, Pentagons, Hexagons, and other figures, instead of remaining fixed in their places, move freely about, on or in the surface, but without the power of rising above or sinking below it, very much like shadows – only hard and with luminous edges – and you will then have a pretty cor-
10 rect notion of my country and countrymen. Alas, a few years ago, I should have said "my universe": but now my mind has been opened to higher views of things.

In such a country, you will perceive at once that it is impossible that there should be anything of what you call a "solid"
15 kind; but I dare say you will suppose that we could at least distinguish by sight the Triangles, Squares, and other figures, moving about as I have described them. On the contrary, we could see nothing of the kind, not at least so as to distinguish one figure from another. Nothing was visible, nor could be
20 visible, to us, except straight Lines; and the necessity of this I will speedily demonstrate.

Place a penny on the middle of one of your tables in Space; and leaning over it, look down upon it. It will appear a circle.

But now, drawing back to the edge of the table, gradually
25 lower your eye (thus bringing yourself more and more into the condition of the inhabitants of Flatland), and you will find

Notes on Section 1.

1.1. I call our world Flatland. The Square never reveals what his countrymen call their land; he has chosen the adjective "flat" for the sake of his readers from three-dimensional space. By "flat" he means without curvature, but he may also intend it to mean dull or monotonous. "Flatland" deviates from the naming convention for the spaces of other dimensions; the term analogous to Pointland, Lineland, and Spaceland is "Planeland." Several authors have illustrated the problem of determining the nature of space with a story of beings confined to a two-dimensional surface. The most fully developed of these stories is found in *The Shape of Space*, Jeffrey Weeks's beautifully written introduction to the basic geometry of two- and three-dimensional space. See also Dionys Burger's *Sphereland: A Fantasy about Curved Spaces and an Expanding Universe*.

1.2. happy. Lucky.

1.3. Space. Three-dimensional space. Abbott's peculiar use of capitals is irregular, but a few guiding principles are apparent. The names of geometric figures are capitalized whenever they represent Flatlanders. Dimension is always capitalized; "truth" is capitalized whenever it means the truth of the existence of higher-dimensional space. Occasionally, capitalization is used to express emphasis or personification, or to call attention to metaphor, but there are manifest inconsistencies. "Space" is capitalized in 51 of the 54 instances where it means one-, two-, or three-dimensional space; there is no obvious reason for the exceptions.

1.4. straight Lines. The Square consistently uses "straight line" for what is now called "line segment" and "line" for what is now called "curve," the trace of a moving point. His usage is consistent with Euclid's *Elements*, where a "straight line" is infinite only in the sense that it may be extended indefinitely.

1.7. on or in the surface. For an illustration of the important distinction between "on a surface" and "in a surface," see Weeks's account of the "Universal Survey of all of Flatland" in which A Square and his fellow surveyors (who live in their two-dimensional universe) all return to Flatsburgh as their own mirror images. Such a reversal of orientation could not be achieved by a figure that remains on a surface (Weeks 2002, 3–9; 45–49; 65–69).

1.8. like shadows. Perhaps an allusion to Sophocles' *Ajax* ("Phantoms, all we that live, mere fleeting shadows.") or 1 Chronicles 29:15 ("Our days on the earth are as a shadow"). In any case, the representation of humans as two-dimensional figures symbolizes the insubstantiality of human existence.

1.9. luminous. The *Oxford English Dictionary* credits Charles Darwin with being the first person to use "luminous" to describe animals or plants that emit light. In Flatland, the identification of both animate and inanimate objects from their appearance depends upon the luminosity of their perimeters.

1.25. gradually lower your eye. A similar thought experiment appeared in an essay, "On space of four dimensions," which appeared in the 1 May 1873 issue of *Nature* (Rodwell 1873). This essay is a revision of a lecture that George F. Rodwell delivered at a meeting of the Natural History Society at Marlborough

the penny becoming more and more oval to your view; and
at last when you have placed your eye exactly on the edge
of the table (so that you are, as it were, actually a Flatlander)
30   the penny will then have ceased to appear oval at all, and
will have become, so far as you can see, a straight line.

The same thing would happen if you
were to treat in the same way a Triangle,
or Square, or any other figure cut out of
35   pasteboard. As soon as you look at it with
your eye on the edge on the table, you
will find that it ceases to appear to you a
figure, and that it becomes in appearance
a straight line. Take for example an equi-
40   lateral Triangle – who represents with us
a Tradesman of the respectable class.
Fig. 1 represents the Tradesman as you
would see him while you were bending
over him from above; figs. 2 and 3 repre-
45   sent the Tradesman, as you would see him if your eye were
close to the level, or all but on the level of the table; and if your
eye were quite on the level of the table (and that is how we see
him in Flatland) you would see nothing but a straight line.

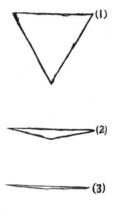

When I was in Spaceland I heard that your sailors have
50   very similar experiences while they traverse your seas and
discern some distant island or coast lying on the horizon.
The far-off land may have bays, forelands, angles in and out
to any number and extent; yet at a distance you see none
of these (unless indeed your sun shines bright upon them
55   revealing the projections and retirements by means of light
and shade), nothing but a grey unbroken line upon the water.

Well, that is just what we see when one of our triangular or
other acquaintances comes towards us in Flatland. As there
is neither sun with us, nor any light of such a kind as to
60   make shadows, we have none of the helps to the sight that
you have in Spaceland. If our friend comes close to us we see

College, where he was the science master. That a prestigious journal like *Nature* published this essay, which is addressed to a general audience, indicates the extensive popular interest in the fourth dimension in the late nineteenth century.

1.27. oval. A penny viewed obliquely appears to be an elliptical disc, in popular language, an oval. An ellipse is a closed curve formed by the intersection of a right circular cone and a plane.

Figure 1.1. An old English penny "becoming more and more oval."

1.29. Flatlander. The first edition reads "Flatland citizen."

1.44. Figures. The crudely drawn illustrations in *Flatland*, described on the title page as "by the Author, A Square," are one of several means that Abbott employs to give his text an ancient character.

1.49. When I was in Spaceland. As we shall see in Part II, the Square has visited Spaceland, our three-dimensional space.

   We noted in the introduction that there are a number of parallels between *Flatland* and Plato's parable of the cave. One significant difference between these stories is their points of view. Plato's story is told in the third person by Socrates; *Flatland* is Plato's parable told as the first-person account of a prisoner who has escaped the cave and returned.

1.49. your sailors. For Abbott's brother Edward, a sailor who died at sea, see Appendix B1, 1859.

1.60. nor any light of such a kind as to make shadows. In Flatland, there is a diffuse light that provides uniform illumination and so does not produce highlights or shadows, which would give visual clues.

1.61. If our friend. Many sentences in *Flatland* contain alliteration, but none contains more than the following specimen, which ends with an "else" to flag the parade of "l's":

> If our friend comes closer to us we see his line becomes larger; if he leaves us it becomes smaller: but still he looks like a straight line; be he a Triangle, Square, Pentagon, Hexagon, Circle, what you will – a straight line he looks and nothing **else**.

The repeated "l's" do not constitute alliteration in its current, most restrictive sense: the repetition of an initial consonant sound in two or more neighboring words or syllables. However, Abbott and Seeley's definition of alliteration includes what they call "concealed alliteration" – cases where the alliteration depends not upon the initial but upon the middle syllables of words or where the alliterative words are separated from one another. There is a good deal of this type of "alliteration" throughout *Flatland*, even though Abbott and Seeley

his line becomes larger; if he leaves us it becomes smaller:
but still he looks like a straight line; be he a Triangle, Square,
Pentagon, Hexagon, Circle, what you will – a straight Line
65   he looks and nothing else.

You may perhaps ask how under these disadvantageous
circumstances we are able to distinguish our friends from
one another: but the answer to this very natural question
will be more fitly and easily given when I come to describe
70   the inhabitants of Flatland. For the present let me defer this
subject, and say a word or two about the climate and houses
in our country.

§2
Of the Climate and Houses
in Flatland

AS WITH YOU, SO ALSO with us, there are four points of
the compass North, South, East, and West.

There being no sun nor other heavenly bodies, it is impossible for us to determine the North in the usual way; but we
5   have a method of our own. By a Law of Nature with us, there
is a constant attraction to the South; and, although in temperate climates this is very slight – so that even a Woman in
reasonable health can journey several furlongs northward
without much difficulty – yet the hampering effect of the
10   southward attraction is quite sufficient to serve as a compass
in most parts of our earth. Moreover the rain (which falls at
stated intervals) coming always from the North, is an additional assistance; and in the towns we have the guidance of
the houses, which of course have their side-walls running
15   for the most part North and South, so that the roofs may
keep off the rain from the North. In the country, where there
are no houses, the trunks of the trees serve as some sort of

caution against its overuse: "Alliteration was from the earliest times noticed by the English ear. By itself, without rhyme, it was once sufficient to constitute poetry... This may explain why an excess of alliteration in prose is peculiarly offensive" (Abbott and Seeley 1871, 97, 185–186).

Alliteration is one of several deliberate archaisms that Abbott uses to convey remoteness of time and place. Indeed, he may have intended the disguised alliteration as a reminiscence of Latin verse. As a student at the University of Cambridge in 1860, Abbott won the Camden Medal, awarded for the best poem in Latin hexameter verse.

1.64. Circle. Circular disc.

Notes on Section Two

2.8. furlong. One-eighth of a mile, or 220 yards.

2.12. stated intervals. Regular in occurrence. At line 15.3, we learn that Flatlanders use the regular occurrence of rain to measure the passing of time.

guide. Altogether, we have not so much difficulty as might be expected in determining our bearings.

20   Yet in our more temperate regions, in which the southward attraction is hardly felt, walking sometimes in a perfectly desolate plain where there have been no houses nor trees to guide me, I have been occasionally compelled to remain stationary for hours together, waiting till the rain came before 25 continuing my journey. On the weak and aged, and especially on delicate Females, the force of attraction tells much more heavily than on the robust of the Male Sex, so that it is a point of breeding, if you meet a Lady in the street always to give her the North side of the way – by no means an easy 30 thing to do always at short notice when you are in rude health and in a climate where it is difficult to tell your North from your South.

Windows there are none in our houses: for the light comes to us alike in our homes and out of them, by day and by night, 35 equally at all times and in all places, whence we know not. It was in old days, with our learned men, an interesting and oft-investigated question, "What is the origin of light?" and the solution of it has been repeatedly attempted, with no other result than to crowd our lunatic asylums with the would- 40 be solvers. Hence, after fruitless attempts to suppress such investigations indirectly by making them liable to a heavy tax, the Legislature, in comparatively recent times, absolutely prohibited them. I – alas, I alone in Flatland – know now only too well the true solution of this mysterious prob- 45 lem; but my knowledge cannot be made intelligible to a single one of my countrymen; and I am mocked at – I, the sole possessor of the truths of Space and of the theory of the introduction of Light from the world of three Dimensions – as if I were the maddest of the mad! But a truce to these 50 painful digressions: let me return to our houses.

2.18. we have not so much difficulty. In modern (and Victorian) usage, one would use the auxiliary *do*: "we do not have so much difficulty."

To give *Flatland* the impression of an antiquated style, Abbott employs what has been called "negative archaism," that is, he creates a time illusion with a sparing use of archaic words and constructions and by avoiding modes of expression that are essentially modern (Fowler and Fowler 1906).

2.24. for hours together. Without intermission.

2.31. rude health. Robust or vigorous health.

2.33. Windows there are none. Although Flatlanders do not have windows in their homes, they do have "glass," which they have used to make half-hour glasses.

2.37. What is the origin of light? Like the prisoners in Plato's cave, Flatlanders do not know the origin of light.

2.42. heavy tax. A Flatland version of the English "taxes on knowledge," a highly refined means of censorship first imposed in 1712 in the form of stamp duties on paper, pamphlets, books, and newspapers. They were not completely abolished until 1861.

2.49. a truce to. Enough of.

The most common form for the construction of a house is five-sided or pentagonal, as in the annexed figure. The two Northern sides *RO, OF*, constitute the roof, and for the most part have no doors; on the East is
55 a small door for the Women; on the West a much larger one for the Men; the South side or floor is usually doorless.

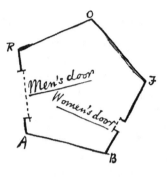

Square and triangular houses
60 are not allowed, and for this reason. The angles of a Square (and still more those of an equilateral Triangle) being much more pointed than those of a Pen-
65 tagon, and the lines of inanimate objects (such as houses) being dimmer than the lines of Men and Women, it follows that there is no little danger lest the points of a square or triangular house residence might do serious injury to an inconsiderate or perhaps absent-minded traveller suddenly
70 running against them: and therefore, as early as the eleventh century of our era, triangular houses were universally forbidden by Law, the only exceptions being fortifications, powder-magazines, barracks, and other state buildings, which it is not desirable that the general public should approach with-
75 out circumspection.

At this period, square houses were still everywhere permitted, though discouraged by a special tax. But, about three centuries afterwards, the Law decided that in all towns containing a population above ten thousand, the angle of a
80 Pentagon was the smallest house-angle that could be allowed consistently with the public safety. The good sense of the community has seconded the efforts of the Legislature; and now, even in the country, the pentagonal construction has superseded every other. It is only now and then in some
85 very remote and backward agricultural district that an antiquarian may still discover a square house.

2.53. sides RO, OF, constitute the roof. Note the playful labeling of the roof.

2.59. diagram of house. For the reader, this figure is a diagram or blueprint, but for a Flatlander, it would be a scale model. The Square does not explain what keeps his house "together," but interior supports and a series of "locks" would permit entry and exit while ensuring that the configuration did not collapse.

2.59. triangular houses. Sir Thomas Tresham built the extraordinary Rushton Triangular Lodge in Northamptonshire on his return in 1593 from imprisonment for his Roman Catholic beliefs. The lodge symbolizes the Holy Trinity – it has three sides, three floors, trefoil windows, and three triangular gables on each side.

In *A Home for All* (1853), Orson Fowler maintained the superiority of octagonal houses over rectangular and square structures. According to Fowler, an octagonal house was cheaper to build, allowed for additional living space, received more natural light, and was easier to heat in winter and keep cool in summer. As a result of Fowler's book, several thousand octagonal houses were built in the United States, mainly on the East Coast and in the Midwest.

2.75. circumspection. This is one of many instances of geometric wordplay in *Flatland*. Circumspection is intended both in the literal sense, looking around the perimeter, as well as the sense of vigilant and cautious observation.

2.77. special tax. Introduced in 1696, the House and Window Tax assessed householders "according to the size and grandeur of their properties" (Douglas 1999, 15).

2.86. antiquarian. The Victorians were fascinated with the past, and local antiquarian societies proliferated during the nineteenth century. Antiquarians studied a wide range of antiquities, including manuscripts, coins, pottery, statues, and buildings.

§3
Concerning the Inhabitants
of Flatland

THE GREATEST LENGTH OR BREADTH of a full-grown inhabitant of Flatland may be estimated at about eleven of your
inches. Twelve inches may be regarded as a maximum.

Our Women are Straight Lines.

5   Our Soldiers and Lowest Classes of Workmen are Triangles with two equal sides, each about eleven inches long,
and a base or third side so short (often not exceeding half
an inch) that they form at their vertices a very sharp and
formidable angle. Indeed when their bases are of the most
10   degraded type (not more than the eighth part of an inch in
size), they can hardly be distinguished from Straight Lines
or Women; so extremely pointed are their vertices. With us,
as with you, these Triangles are distinguished from others
by being called Isosceles; and by this name I shall refer to
15   them in the following pages.

Our Middle Class consists of Equilateral or Equal-sided
Triangles.

Our Professional Men and Gentlemen are Squares (to
which class I myself belong) and Five-sided Figures or
20   Pentagons.

Next above these come the Nobility, of whom there are several degrees, beginning at Six-sided Figures, or Hexagons,
and from thence rising in the number of their sides till

Notes on Section 3.

3.1. length or breadth. The Square uses the words "length" and "breadth" in a way that may not be consistent with the reader's understanding of these words; he does not use the word "width." Informally, a Flatlander's length is the width of the widest hallway through which he or she can block passage. [continued]

3.3. Twelve inches . . . maximum. The Russian scientist and poet, Nicholas Morosoff, was imprisoned for more than twenty years in the fortress of Schlüsselburg ("the Russian Bastille") for his revolutionary activities. In a letter written to his fellow prisoners in 1891, he illustrates the problem of imagining four-dimensional space with a description of two-dimensional beings floating on the surface of a lake, and how they would perceive a human being who entered the water. [continued]

3.4. Women are Straight Lines. Flatland women are mere (one-dimensional) line segments, a fitting representation of their relegation to the narrowly defined role as child-bearers and housekeepers.

3.9. Soldiers . . . very sharp and formidable angle. Abbott says in *The Kernel and the Husk*, "I honour the army as much as most men, more perhaps than many do: but after all the profession of a soldier is the profession of a throat-cutter; throat-cutting in an extensive, expeditious, and honourable way – throat-cutting in one direction often undertaken merely to prevent throat-cutting in another direction – but still throat-cutting after all" (Abbott 1886, 60).

3.14. Isosceles. In Flatland, a triangle is said to be isosceles provided that it has exactly two sides of equal length.

3.16. Class. Flatland's hierarchical society parodies the rigorously stratified society of nineteenth-century England. In his essay, "London at midsummer" (1877), the novelist Henry James observed, "The essentially hierarchical plan of English society is the great and ever-present fact. . . There is hardly a detail of life that does not in some degree betray it" (James 1905, 158).

3.18. Gentlemen. The definition of a Flatland "gentleman" is consistent with the historic definition of an English gentleman: one who is lawfully entitled to armorial distinction (i.e., a coat of arms with some mark of distinction) but is not included in any recognized degree of nobility (Cussans 1893, 215).

3.19. Squares (to which class I myself belong). Abbott and Seeley's analysis of *Robinson Crusoe* provides a fitting description of *Flatland*. They observe that its narrator has no peculiarity of character that differentiates him from other people. The story is therefore not properly a study of character but a study of human nature in general, and an exceptional incident is necessary to make it interesting (Abbott and Seeley 1871, 254).

they receive the honourable title of Polygonal, or many-
sided. Finally when the number of the sides becomes so
numerous, and the sides themselves so small, that the figure
cannot be distinguished from a circle, he is included in the
Circular or Priestly order; and this is the highest class of all.

It is a Law of Nature with us that a male child shall have
one more side than his father, so that each generation shall
rise (as a rule) one step in the scale of development and
nobility. Thus the son of a Square is a Pentagon; the son of a
Pentagon, a Hexagon; and so on.

But this rule applies, not always to the Tradesmen, and still
less often to the Soldiers, and to the Workmen; who indeed
can hardly be said to deserve the name of human Figures,
since they have not all their sides equal. With them therefore
the Law of Nature does not hold; and the son of an Isosceles
(*i.e.* a Triangle with two sides equal) remains Isosceles still.
Nevertheless, all hope is not shut out, even from the Isosceles, that his posterity may ultimately rise above his degraded
condition. For, after a long series of military successes, or dili-
gent and skilful labours, it is generally found that the more
intelligent among the Artisan and Soldier classes manifest a
slight increase of their third side or base, and a shrinkage of
the two other sides. Intermarriages (arranged by the Priests)
between the sons and daughters of these more intellectual
members of the lower classes generally result in an offspring
approximating still more to the type of the Equal-sided
Triangle.

Rarely – in proportion to the vast numbers of Isosceles
births – is a genuine and certifiable Equal-sided Triangle
produced from Isosceles parents.[1] Such a birth requires,

---

[1] "What need of a certificate?" a Spaceland critic may ask: "Is not the
procreation of a Square Son a certificate from Nature herself, proving the
Equal-sidedness of the Father?" I reply that no Lady of any position will
marry an uncertified Triangle. Square offspring has sometimes resulted
from a slightly Irregular Triangle: but in almost every such case the
Irregularity of the first generation is visited on the third; which either
fails to attain the Pentagonal rank, or relapses to the Triangular.

3.24. Polygonal. A polygon is a plane figure bounded by a finite number of line segments (called sides); the word is derived from the Greek *polus* "many" and *gonia* "angle." Although the Square defines polygonal as "many-sided," it literally means having many angles. In Flatland, all polygons are convex; that is, for every pair of points P and Q inside the polygon, the line segment that joins P and Q lies within the polygon. In these notes, "polygon" means "convex polygon."

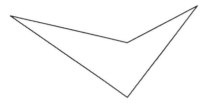

Figure 3.2. A non-convex quadrilateral.

3.28. Circular or Priestly order. In Flatland, "Priest" is not a religious title; the Circles, or Priests, are just the ruling class. Abbott may have gotten the idea for this extraordinary usage of "priest" from his friend John R. Seeley, who compared the Church to a state and the priest to a statesman in his controversial book, *Ecce Homo.*

3.28. highest class of all. A circle is a traditional symbol for perfection.

3.29. Law of Nature. This "evolutionary law" is akin to a theory of Jean-Baptiste Lamarck (1744–1829), the French biologist, who regarded evolution (which he called "transformism") as a natural process of increasing complexity tending toward "perfection." [continued]

3.36. hardly . . . human. Aristotle expressed the judgment that is behind slavery in all its various forms: the enslaved are not fully human (*Politics*, Book I).

3.46. Intermarriage. Intermarriage has two opposite meanings: marriage between persons belonging to different tribes or groups (exogamy) or, as here, marriage between persons of the same tribe or kin group (endogamy). [continued]

3.48. result in an offspring. The doctrine that characteristics acquired during an organism's lifetime are inherited by its offspring ("use inheritance") was held almost universally from antiquity. Lamarck was the first person to set forth in detail a theory that evolution occurs through the mechanism of the inheritance by progeny of characteristics acquired by their ancestors in direct response to environmental conditions. [continued]

3.53. Isosceles parents. By an "isosceles mother," the Square means a woman with an isosceles father.

as its antecedents, not only a series of carefully arranged
55   intermarriages, but also a long-continued exercise of frugal-
ity and self-control on the part of the would-be ancestors of
the coming Equilateral, and a patient, systematic, and continuous development of the Isosceles intellect through many
generations.

60   The birth of a True Equilateral Triangle from Isosceles parents is the subject of rejoicing in our country for many furlongs round. After a strict examination conducted by the
Sanitary and Social Board, the infant, if certified as Regular,
is with solemn ceremonial admitted into the class of Equilat-
65   erals. He is then immediately taken from his proud yet sor-
rowing parents and adopted by some childless Equilateral,
who is bound by oath never to permit the child henceforth
to enter his former home or so much as to look upon his
relations again, for fear lest the freshly developed organism
70   may, by force of unconscious imitation, fall back again into
his hereditary level.

The occasional emergence of an Isosceles from the ranks
of his serf-born ancestors, is welcomed not only by the poor
serfs themselves, as a gleam of light and hope shed upon
75   the monotonous squalor of their existence, but also by the
Aristocracy at large; for all the higher classes are well aware
that these rare phenomena, while they do little or nothing to
vulgarise their own privileges, serve as a most useful barrier
against revolution from below.

80   Had the acute-angled rabble been all, without exception,
absolutely destitute of hope and of ambition, they might
have found leaders in some of their many seditious outbreaks, so able as to render their superior numbers and
strength too much for the wisdom even of the Circles. But
85   a wise ordinance of Nature has decreed that, in proportion
as the working-classes increase in intelligence, knowledge,
and all virtue, in that same proportion their acute angle
(which makes them physically terrible) shall increase also
and approximate to the comparatively harmless angle of the
90   Equilateral Triangle. Thus, in the most brutal and formidable

3.66. adopted by some childless. The classical Greeks dreaded "childlessness" (the absence of legitimate male offspring), and adoption was a common practice. Legal adoption of children was not possible in England until 1926.

3.67. bound by oath. The oath played an important role in the law and life of the ancient Greeks. Oaths were required of signatories to treaties, of parties to legal disputes, commercial and private contracts, conspiracies, and marriages (Plescia 1970, 3).

3.70. unconscious imitation. Walter Bagehot argued that the stability and order of a social group originates in the human propensity to imitate, which he characterized as instinctive and unconscious. "This unconscious imitation and encouragement of appreciated character, and the equally unconscious shrinking from and persecution of disliked character, is the main force which moulds and fashions men in society as we now see it." In Bagehot's (Lamarckian) derivation of social organization, a tradition first acquired by unconscious imitation becomes an instinct by use inheritance (Bagehot 1872, 97).

3.72. emergence of an Isosceles. What Abbott means is "emergence of an Isosceles from his former station into the ranks of the Equilaterals." In the edition published by Basil Blackwell in 1926, "Isosceles" was changed to "Equilateral," and subsequent editions have retained this "correction."

3.73. ranks of his serf-born. The Isosceles triangles are said to be serfs. A serf is a person in a condition of servitude who is bound to a hereditary plot of land and to the will of his master. Serfdom differs from slavery in that the services due to the master and his power of disposal of his serf are limited by law or custom. It had virtually disappeared from England by the end of the fifteenth century. [continued]

3.76. Aristocracy. In the literal sense of the Greek, an aristocracy is the rule of a state by its best citizens; Plato advocates rule by an aristocracy of intellect. Flatland is ruled by an aristocracy of birth. In the nineteenth century, England was dominated socially, economically, and politically by an aristocracy comprised of the great landed proprietors, who were generally hereditary nobles.

3.78. vulgarise. Victorian authorities differed on the question whether verbs like "vulgarise" and nouns like "civilisation" should be spelled with an *s* or a *z* (Fowler and Gowers 1965, 314). Abbott invariably uses *-ise* in his other writings. The Square is not even consistent with the same word; for example, he uses "recognize" twenty times and "recognise" three times.

3.82. seditious. Engaged in inciting a revolt against constituted authority.

3.84. too much for the wisdom even of the Circles. The first edition reads, "too much even for the wisdom of the Circles."

of the soldier class creatures almost on a level with women in their lack of intelligence – it is found that, as they wax in the mental ability necessary to employ their tremendous penetrating power to advantage, so do they wane in the
95   power of penetration itself.

How admirable is this Law of Compensation! And how perfect a proof of the natural fitness and, I may almost say, the divine origin of the aristocratic constitution of the States in Flatland! By a judicious use of this Law of Nature, the
100 Polygons and Circles are almost always able to stifle sedition in its very cradle, taking advantage of the irrepressible and boundless hopefulness of the human mind. Art also comes to the aid of Law and Order. It is generally found possible – by a little artificial compression or expansion on the part of
105 the State physicians – to make some of the more intelligent leaders of a rebellion perfectly Regular, and to admit them at once into the privileged classes; a much larger number, who are still below the standard, allured by the prospect of being ultimately ennobled, are induced to enter the State
110 Hospitals, where they are kept in honourable confinement for life; one or two alone of the more obstinate, foolish, and hopelessly irregular are led to execution.

Then the wretched rabble of the Isosceles, planless and leaderless, are either transfixed without resistance by the
115 small body of their brethren whom the Chief Circle keeps in pay for emergencies of this kind; or else more often, by means of jealousies and suspicions skilfully fomented among them by the Circular party, they are stirred to mutual warfare, and perish by one another's angles. No less than one hundred
120 and twenty rebellions are recorded in our annals, besides minor outbreaks numbered at two hundred and thirty-five; and they have all ended thus.

3.96. How admirable is this Law of Compensation! William Paley's influential *Natural Theology* (1802) is best known for its presentation of the argument from design in the form of an analogy. Just as we infer the existence of a watchmaker from the existence of a watch, so can we infer the existence of an intelligent creator from the character of objects in the natural world.

Chapter XVI (Compensation) of *Natural Theology* opens with an argument that an elephant's trunk is the product of design rather than (Lamarckian) evolution. According to Paley, the creator gave the elephant a short unbending neck to accommodate the weight of its head and then compensated for such a neck by giving it a long, flexible proboscis (Paley 1802, 147).

3.98. divine origin of the aristocratic constitution. Flatland's "constitution" is not a civil constitution but a set of natural laws. Abbott is satirizing the argument from natural theology that the social order, like the order of the physical universe, is governed by divine ordinance.

3.110. honourable confinement. Simple detention or custody without punishment; sometimes called *custodia honesta*.

3.113. wretched rabble. A doubly contemptuous phrase for the lowest class. Cicero called the common laborers, artisans, and shopkeepers the "wretched starveling rabble" (*misera ac ieiuna plebecula*).

Benjamin Disraeli (Victorian novelist and later Prime Minister) described the alienation between England's rich and poor in one of the most famous sentences in Victorian fiction: "Two nations; between whom there is no intercourse and no sympathy; who are as ignorant of each other's habits, thoughts, and feelings, as if they were dwellers in different zones, or inhabitants of different planets; who are formed by a different breeding, are fed by a different food, are ordered by different manners, and are not governed by the same laws" (Disraeli 1845, 67).

3.114. transfix. To pierce through with, or impale upon, a sharp-pointed instrument.

3.115. brethren. A plural form of brother, used chiefly in formal address. Here the word means "fellow-Isosceles."

3.120. rebellions. The ever-present threat of a rebellion by the Flatland serfs has a parallel in ancient Sparta, where political thought was deeply concerned with avoiding a revolution by the helots, who greatly outnumbered the Spartan citizens.

Although England did not experience a massive uprising like the French Revolution, social and economic turmoil kept the Victorians fearful of a revolution throughout the first half of the nineteenth century. Chartism, a movement that demanded political rights for working-class men, posed the last revolutionary threat in England; it collapsed in 1848.

3.120. annals. Historical records.

§4
Concerning the Women

IF OUR HIGHLY POINTED TRIANGLES of the Soldier class
are formidable, it may be readily inferred that far more
formidable are our Women. For if a Soldier is a wedge, a
Woman is a needle; being, so to speak, *all* point, at least at
5   the two extremities. Add to this the power of making herself practically invisible at will, and you will perceive that
a Female in Flatland, is a creature by no means to be trifled
with.

But here, perhaps, some of my younger Readers may ask
10   *how* a woman in Flatland can make herself invisible. This
ought, I think, to be apparent without any explanation. However, a few words will make it clear to the most unreflecting.

Place a needle on a table. Then, with your eye on the level
of the table, look at it side-ways, and you see the whole
15   length of it; but look at it end-ways, and you see nothing but
a point: it has become practically invisible. Just so is it with
one of our Women. When her side is turned towards us, we
see her as a straight line; when the end containing her eye
or mouth – for with us these two organs are identical – is the
20   part that meets our eye, then we see nothing but a highly
lustrous point; but when the back is presented to our view,
then – being only sub-lustrous, and, indeed, almost as dim
as an inanimate object – her hinder extremity serves her as a
kind of Invisible Cap.

25   The dangers to which we are exposed from our Women
must now be manifest to the meanest capacity in Spaceland.
If even the angle of a respectable Triangle in the middle class
is not without its dangers; if to run against a Working Man
involves a gash; if collision with an Officer of the military

Notes on Section 4.

4.4. Woman is a needle. Perhaps an allusion to these well-known lines from Tennyson's poem, "The Princess" (1849):

> Man for the field and woman for the hearth:
> Man for the sword and for the needle she:

For a middle-class Victorian woman, proficiency at needlework and embroidery was a traditional means of demonstrating her fitness for marriage and motherhood.

4.6. invisible. Nineteenth-century English wives were "invisible" before the law. In his famous *Commentaries* (1765), William Blackstone described the legal doctrine of coverture, according to which the husband and wife were as one and that one was the husband. Coverture was only partially amended by the Married Women's Property Acts of 1879, 1882, and 1893.

Because females in classical Athens were considered incapable of deciding their own affairs, they were subjected to a practice even more oppressive than coverture. Throughout their lives they were under the dominion of a *kyrios* (a male "lord" or "guardian"), who was their source of support and served as their representative in legal transactions. The *kyrios* of a girl (normally her father) had the right to decide whom she would marry, and after she married, her husband became her *kyrios* (Gagarin and Cohen 2005, 245–246).

4.8. trifled with. As a needle, a woman is at once an object of trifling importance and a creature not to be trifled with.

4.24. Invisible Cap. In Greek mythology, the cap of invisibility (also called the cap of Hades) is a cap that renders its wearers invisible; for example, Athena uses it in the *Iliad.*

30   class necessitates a serious wound; if a mere touch from the vertex of a Private Soldier brings with it danger of death; – what can it be to run against a Woman, except absolute and immediate destruction? And when a Woman is invisible, or visible only as a dim sub-lustrous point, how difficult must
35   it be, even for the most cautious, always to avoid collision!

Many are the enactments made at different times in the different States of Flatland, in order to minimize this peril; and in the Southern and less temperate climates where the force of gravitation is greater, and human beings more
40   liable to casual and involuntary motions, the Laws concerning Women are naturally much more stringent. But a general view of the Code may be obtained from the following summary:–

1 Every house shall have one entrance in the Eastern side,
45   for the use of Females only; by which all females shall enter "in a becoming and respectful manner"[2] and not by the Men's or Western door.
2 No Female shall walk in any public place without continually keeping up her Peace-cry, under penalty of
50   death.
3 Any Female, duly certified to be suffering from St. Vitus's Dance, fits, chronic cold accompanied by violent sneezing, or any disease necessitating involuntary motions, shall be instantly destroyed.

55   In some of the States there is an additional Law forbidding Females, under penalty of death, from walking or standing in any public place without moving their backs constantly from right to left so as to indicate their presence to those behind them; others oblige a Woman, when travelling, to be followed
60   by one of her sons, or servants, or by her husband; others confine Women altogether to their houses except during the

---

[2] When I was in Spaceland I understood that some of your Priestly Circles have in the same way a separate entrance for Villagers, Farmers, and Teachers of Board Schools (*Spectator*, Sept. 1884, p. 1255) that they may "approach in a becoming and respectful manner."

4.45. entrance . . . Females only. In Athenian households, females lived and worked in separate quarters located in a remote part of the house.

4.49. Peace-cry. Abbott, who read widely, may have been aware of Longfellow's poem, "The Nun of Nidaros," which contains the lines:

> Love against hatred,
> Peace-cry for war-cry!

4.52. St. Vitus's Dance. A convulsive disorder characterized by irregular involuntary contractions of the muscles; also known as chorea.

4.56. standing. In a footnote in Section 15, the Square explains that "lying," "sitting," and "standing" in Flatland are mental states of volition.

4.61. confine Women. The seclusion of Flatland women has a parallel in classical Athens, where women were supposed to keep out of public view, with the certain exceptions including appearances at funerals and festivals.

Footnote 2. The footnote refers to an item reproduced below, which appeared in *The Spectator* (27 September 1884, 1255). The submerged message of this odd footnote is a comment on the arrogance of certain English clergy, not the similarity between the women's entrance to Flatland homes and the backdoor entrance for villagers at the South Wytham Rectory.

A worthy correspondent is angry with a Missionary Society which receives converts in India without compelling them to break their caste. The Society is probably unwise, though caste is not exactly what Mr. Dyson thinks; but what will he say to the following letter, addressed to the schoolmaster, South Wytham, by the rector of that parish, and published in the *Stamford Mercury*:–

South Wytham Rectory, September 17th, 1884.

All the villagers desirous of coming to my house approach it in a becoming and respectful manner, –*i.e.*, through the backway and to the kitchen door. There is not a farmer in the place who ever has had, or would have, the impertinence to do otherwise. I desire that in the future you will do the like.

(Signed),
R. W. L. Tollemache-Tollemache
The Teacher of the Board School, South Wytham.

It is that spirit, and not any dissatisfaction either with its dogmas or its ritual, which, when Disestablishment is proposed, may endanger the Church of England. It will be observed that the exclusion is not in any way personal to the school-master, who ought to rank, if there is to be rank, next after the cleric, but extends to five-sixths of Mr. Tollemache's congregation.

religious festivals. But it has been found by the wisest of our
Circles or Statesmen that the multiplication of restrictions on
Females tends not only to the debilitation and diminution of
65   the race, but also to the increase of domestic murders to
such an extent that a State loses more than it gains by a too
prohibitive Code.

For whenever the temper of the Women is thus exasperated
by confinement at home or hampering regulations abroad,
70   they are apt to vent their spleen upon their husbands and
children; and in the less temperate climates the whole male
population of a village has been sometimes destroyed in one
or two hours of simultaneous female outbreak. Hence the
Three Laws, mentioned above, suffice for the better regulated
75   States, and may be accepted as a rough exemplification of our
Female Code.

After all, our principal safeguard is found, not in Legislature, but in the interests of the Women themselves. For,
although they can inflict instantaneous death by a retrograde
80   movement, yet unless they can at once disengage their sting-
ing extremity from the struggling body of their victim, their
own frail bodies are liable to be shattered.

The power of Fashion is also on our side. I pointed out that
in some less civilised States no female is suffered to stand in
85   any public place without swaying her back from right to
left. This practice has been universal among ladies of any
pretensions to breeding in all well-governed States, as far
back as the memory of Figures can reach. It is considered a
disgrace to any State that legislation should have to enforce
90   what ought to be, and is in every respectable female, a natural
instinct. The rhythmical and, if I may so say, well-modulated
undulation of the back in our ladies of Circular rank is envied
and imitated by the wife of a common Equilateral, who can
achieve nothing beyond a mere monotonous swing, like the
95   ticking of a pendulum; and the regular tick of the Equilateral
is no less admired and copied by the wife of the progressive
and aspiring Isosceles, in the females of whose family no
"back-motion" of any kind has become as yet a necessity

4.62. religious festivals. This is the only place that Flatlanders' religion is mentioned. The most prominent aspect of classical Greek religion was the celebration of public festivals, and we may conjecture that such is the nature of religion in Flatland.

In different ways, religion permeated life in classical Greece and Victorian England alike, and it provided a ready source of figurative language for both Plato and Abbott. Throughout his dialogues, Plato makes mention of gods, festivals, beliefs, and rites. Abbott uses a good deal of "religious" language as well as biblical images and allusions in *Flatland*.

Abbott believed that the cultivation of the intellect increases one's religious faith. In particular, he maintained that the works of Plato, Shakespeare, Francis Bacon, George Eliot, and William Wordsworth, among others, provide invaluable commentary on the Bible. He intended *Flatland* as an exposition of contemporary social and religious issues, and he cites *Flatland* in three subsequent theological volumes, *The Kernel and the Husk*, *The Spirit on the Waters*, and *Apologia*. The first exploration of the relevance of *Flatland* to religious questions was the essay, "Abbott's *Flatland*: Scientific imagination and 'natural Christianity,'" in which Rosemary Jann makes a strong case for reading *Flatland* "as a rebuke to the rigid literalism of both materialist science and fundamentalist religion" (Jann 1985, 478).

4.68. the temper of the Women. One argument for denying women the right to vote was that if they were granted suffrage, "their natural eagerness and quickness of temper would probably make them hotter partisans than men" (Ward 1889, 783).

4.69. abroad. Out of one's house.

4.70. vent their spleen. Let loose their anger.

4.84. suffer. To allow (a thing) to be done or take place.

4.92. well-modulated undulation. In the 1880s, women wore a stuffed pad or small wire framework called a bustle (a "dress-improver") beneath the skirt of their dresses to expand and support them behind. The bustle focused attention on a woman's backside and emphasized the movement of that body part.

of life. Hence, in every family of position and consideration,
100   "back motion" is as prevalent as time itself; and the husbands
and sons in these households enjoy immunity at least from
invisible attacks.

Not that it must be for a moment supposed that our Women
are destitute of affection. But unfortunately the passion of the
105   moment predominates, in the Frail Sex, over every other consideration. This is, of course, a necessity arising from their
unfortunate conformation. For as they have no pretensions
to an angle, being inferior in this respect to the very lowest of the Isosceles, they are consequently wholly devoid of
110   brain-power, and have neither reflection, judgment nor fore-
thought, and hardly any memory. Hence, in their fits of fury,
they remember no claims and recognise no distinctions. I
have actually known a case where a Woman has exterminated her whole household, and half an hour afterwards,
115   when her rage was over and the fragments swept away, has
asked what has become of her husband and her children!

Obviously then a Woman is not to be irritated as long as
she is in a position where she can turn round. When you have
them in their apartments – which are constructed with a view
120   to denying them that power – you can say and do what you
like; for they are then wholly impotent for mischief, and will
not remember a few minutes hence the incident for which
they may be at this moment threatening you with death,
nor the promises which you may have found it necessary to
125   make in order to pacify their fury.

On the whole we get on pretty smoothly in our domestic
relations, except in the lower strata of the Military Classes.
There the want of tact and discretion on the part of the husbands produces at times indescribable disasters. Relying too
130   much on the offensive weapons of their acute angles instead
of the defensive organs of good sense and seasonable simulations, these reckless creatures too often neglect the pre-
scribed construction of the Women's apartments, or irritate
their wives by ill-advised expressions out of doors, which
135   they refuse immediately to retract. Moreover a blunt and

4.105. the Frail Sex. "Frail" can mean either physically or morally weak. Hamlet's condemnation of women, "Frailty, thy name is woman!" most famously expresses the view that women are morally weak.

4.110. wholly devoid of brain-power. The statement that Flatland women are "wholly devoid of brain-power" is contradicted by several instances of women's behavior that demonstrate considerable intelligence.

In 1845, Elizabeth Barrett confided to Robert Browning that "there is a natural inferiority of mind in women," and few of her contemporaries would have challenged her assertion. As evidence of the inferiority of the female intellect, Victorian men cited the lack of achievements by women in science, the arts, and literature. Of course, to make a meaningful contribution to these fields, a woman would need an education – something denied her because of her inferior intellect.

The evolutionary biologist George J. Romanes argued that the difference in brain weights of men and women (on average, five ounces) was the cause of female mental inferiority. He conceded that the treatment of women in the past had been shameful, but maintained that even under the most favorable conditions it would "take many centuries for heredity to produce the missing five ounces of the female brain" (Romanes 1887, 654–655, 666).

4.114. exterminated her whole household. Compare to Euripides' famous play *Medea*, where Medea kills her own children to punish her husband for abandoning her.

4.132. seasonable simulations. Insincere declarations or promises made at opportune times.

stolid regard for literal truth indisposes them to make those
lavish promises by which the more judicious Circle can in
a moment pacify his consort. The result is massacre; not
however without its advantages, as it eliminates the more
140   brutal and troublesome of the Isosceles; and by many of our
Circles the destructiveness of the Thinner Sex is regarded as
one among many providential arrangements for suppressing
redundant population, and nipping Revolution in the bud.

Yet even in our best regulated and most approximately
145   circular families I cannot say that the ideal of family life is
so high as with you in Spaceland. There is peace, in so far
as the absence of slaughter may be called by that name, but
there is necessarily little harmony of tastes or pursuits; and
the cautious wisdom of the Circles has ensured safety at
150   the cost of domestic comfort. In every Circular or Polygonal
household it has been a habit from time immemorial – and
now has become a kind of instinct among the women of our
higher classes – that the mothers and daughters should constantly keep their eyes and mouths towards their husband
155   and his male friends; and for a lady in a family of distinction
to turn her back upon her husband would be regarded as a
kind of portent, involving loss of *status*. But, as I shall soon
shew, this custom, though it has the advantage of safety, is
not without its disadvantages.

160   In the house of the Working Man or respectable Tradesman – where the wife is allowed to turn her back upon her
husband, while pursuing her household avocations – there
are at least intervals of quiet, when the wife is neither seen
nor heard, except for the humming sound of the continu-
165   ous Peace-cry; but in the homes of the upper classes there
is too often no peace. There the voluble mouth and bright
penetrating eye are ever directed towards the Master of the
household; and light itself is not more persistent than the
stream of feminine discourse. The tact and skill which suf-
170   fice to avert a Woman's sting are unequal to the task of
stopping a Woman's mouth; and as the wife has absolutely
nothing to say, and absolutely no constraint of wit, sense, or
conscience to prevent her from saying it, not a few cynics

4.138. consort. A husband or wife; a spouse. Prince Albert, the husband of Queen Victoria, was called the prince-consort.

The right of one spouse to the society or services (companionship, love, affection, comfort, sexual intercourse) of the other spouse is generally referred to as the right of consortium. Until the courts held otherwise in the case of *Regina v. Jackson* (1891), a husband was legally entitled to use physical force to secure the consortium of his wife.

4.141. the Thinner Sex. A pun on two meanings of thinner sex: the sex of little thickness and the sex that thins, that is, reduces in number. Paley cites "thinnings," in which one species reduces the number of another species, as evidence of providential design (Paley 1802, 249).

4.142. providential arrangement. The ancient Greeks believed that there was a law of balance or compensation in nature as well as human affairs. A natural system of checks and balances is an underlying theme of *The Histories* of Herodotus (fifth century BC). Herodotus holds that divine forethought has made prolific those species that are timid and easy prey to ensure their continuance, and made the savage and noxious species comparatively unproductive.

4.145. best regulated . . . families. An allusion to Dickens's *David Copperfield*, where Mr. Micawber says, "My dear friend Copperfield, accidents will occur in the best-regulated families."

4.145. ideal of family life. Abbott is probably referring to the great difference between the family in classical Greece and the one in the Victorian age.

In ancient Greek, there was no word for "family" as it is now defined; the nearest equivalent is *oikos* (estate or household), which emphasizes property while ignoring affective relationships. There was seldom real companionship between husband and wife, partly because men generally married at around thirty, whereas women married between fourteen and eighteen. The Greeks demanded absolute marital fidelity of the wife, but no such exclusivity was required of the husband (Pomeroy 1994, 31–40).

The Victorian family was often idealized, as in Coventry Patmore's poem *Angel in the House* (1854) and John Ruskin's lecture *Of Queens' Gardens* (1864). In his survey of the Victorian era, George M. Young describes the belief that the family is "a Divine appointment for the comfort and education of mankind" as a "vital article of the common Victorian faith" (Young 1936, 159).

4.152. habit . . . instinct. The Square has a Lamarckian conception of instinct as inherited habit. Robert J. Richards has traced Darwin's journey from a theory in which habits are preserved because of their usefulness (via use inheritance) to one in which individuals are preserved because of their useful habits (via natural selection) (Richards 1987).

4.157. portent. The superstitious Greeks took abnormal events as omens.

4.158. shew. "Shew" is an archaic form of "show." This is one of only two places where the Square uses "shew" rather than "show."

4.162. avocations. Tasks.

4.163. intervals of quiet. This ancient stereotype of women is found in Sophocles' *Ajax*: "Silence is a woman's glory."

4.166. voluble. Rapid and ready of speech.

175 have been found to aver that they prefer the danger of the death-dealing but inaudible sting to the safe sonorousness of a Woman's other end.

To my readers in Spaceland the condition of our Women may seem truly deplorable, and so indeed it is. A Male of the lowest type of the Isosceles may look forward to some
180 improvement of his angle, and to the ultimate elevation of the whole of his degraded caste; but no Woman can entertain such hopes for her sex. "Once a Woman, always a Woman" is a Decree of Nature; and the very Laws of Evolution seem suspended in her disfavour. Yet at least we can admire the
185 wise Prearrangement which has ordained that, as they have no hopes, so they shall have no memory to recall, and no fore-thought to anticipate, the miseries and humiliations which are at once a necessity of their existence and the basis of the constitution of Flatland.

§5
Of our Methods of Recognizing One Another

You, WHO ARE BLESSED WITH shade as well as light, you who are gifted with two eyes, endowed with a knowledge of perspective, and charmed with the enjoyment of various colours, you, who can actually *see* an angle, and contemplate
5 the complete circumference of a Circle in the happy region of the Three Dimensions – how shall I make clear to you the extreme difficulty which we in Flatland experience in recognizing one another's configuration?

4.174. they prefer the danger. The use of a woman's tongue as a weapon is the subject of various English proverbs.

4.175. sonorousness. The quality of giving a sound.

4.184. Laws of Evolution seem suspended in her disfavour. The influential social theorist Herbert Spencer argued that individual evolution is arrested earlier in women than in men to conserve energy for reproduction. Furthermore, this "arrest of individual development" in a woman causes a "perceptible falling-short" in both her capability for abstract reasoning and her sense of justice (Spencer 1873, 374).

4.185. wise Prearrangement. A highly satirical section concludes with the Square's absurd rationalization of the plight of women. Although he acknowledges that their condition is deplorable, it does not occur to him to question the doctrine that the condition of women, like all prevailing social conditions, is divinely ordained. Instead, he expresses his admiration for the wisdom of Providence, which has compensated women by endowing them with so little intelligence that they can neither anticipate nor recall the miseries and humiliations that they daily endure.

Notes on Section 5.

5.3. a knowledge of perspective. It is curious that Flatlanders do not have knowledge of perspective; that is, they do not understand precisely how the appearance of an object depends upon its spatial relationship to the observer. For a human artist, the "problem of perspective" is how to render a three-dimensional object on a two-dimensional surface so that the rendered image gives the same impression of apparent relative position and magnitude as does the object itself. Judith V. Field's excellent *The Invention of Infinity* tells the story of the discovery and solution of this problem by artist/mathematicians of the Italian Renaissance, and how the mathematical aspects of the problem were generalized into what is now called projective geometry, the study of the properties and invariants of geometric figures under projection (Field 1997).

5.3. charmed with. Favored with.

5.5. contemplate the complete. This phrase is an example of homeoteleuton, or near-rhyme, the similarity of endings of adjacent or parallel words. Two other prominent examples are found at lines 6.141 (propriety/society) and 12.13 (queller/color).

To "contemplate" usually means to meditate upon or view mentally, but here it means to observe or behold. Flatlanders cannot contemplate the complete circumference in this sense; unlike Spacelanders looking down upon a circle, they cannot see the entire circle all at once. Indeed, as Flatlanders move around a circle, they can only see somewhat less than half of its circumference at any given time. [continued]

Recall what I told you above. All beings in Flatland, ani-
10   mate or inanimate, no matter what their form, present *to our
view* the same, or nearly the same, appearance, viz. that of
a straight Line. How then can one be distinguished from
another, where all appear the same?

The answer is threefold. The first means of recognition
15   is the sense of hearing; which with us is far more highly
developed than with you, and which enables us not only to
distinguish by the voice our personal friends, but even to
discriminate between different classes, at least so far as con-
cerns the three lowest orders, the Equilateral, the Square, and
20   the Pentagon – for of the Isosceles I take no account. But as
we ascend in the social scale, the process of discriminating
and being discriminated by hearing increases in difficulty,
partly because voices are assimilated, partly because the faculty of voice-discrimination is a plebeian virtue not much
25   developed among the Aristocracy. And wherever there is
any danger of imposture we cannot trust to this method.
Amongst our lowest orders, the vocal organs are developed
to a degree more than correspondent with those of hearing,
so that an Isosceles can easily feign the voice of a Polygon,
30   and, with some training, that of a Circle himself. A second
method is therefore more commonly resorted to.

Feeling is, among our Women and lower classes – about
our upper classes I shall speak presently – the principal test
of recognition, at all events between strangers, and when
35   the question is, not as to the individual, but as to the class.
What therefore "introduction" is among the higher classes in
Spaceland, that the process of "feeling" is with us. "Permit
me to ask you to feel and be felt by my friend Mr. So-and-
so" – is still, among the more old-fashioned of our country
40   gentlemen in districts remote from towns, the customary
formula for a Flatland introduction. But in the towns, and
among men of business, the words "be felt by" are omitted and the sentence is abbreviated to, "Let me ask you
to feel Mr. So-and-so"; although it is assumed, of course,
45   that the "feeling" is to be reciprocal. Among our still more
modern and dashing young gentlemen – who are extremely
averse to superfluous effort and supremely indifferent to the

5.11. viz. An abbreviation of the Latin word *videlicet* ("it is permissible to see"), usually read as "namely" or "that is."

5.15. sense of hearing. It follows from the work of the Dutch physicist Christian Huygens that sound waves move "sharply" through three-dimensional space but not through two-dimensional space. If we are some distance from a gunshot, we hear nothing until the sound waves reach us, when we hear a rapid report followed by silence. But in Flatland, there would be silence then the sound of the shot followed by a kind of unceasing reverberation (Solomon 1992; Morley 1985).

5.18. discriminate between different classes. In the nineteenth century, the proper use of language and particularly accent was a marker of class status. "No saying was ever truer than that good breeding and good education are sooner discovered from the style of speaking, or the language employed in conversation, than from any other means" (Williams 1850, 5).

5.23. are assimilated. Become alike.

5.36. "introduction." The Victorian code of etiquette specified that introductions should never be made indiscriminately. If there was the slightest doubt about how an introduction would be received, then both persons should be asked in advance whether they wished to be introduced. For two persons of different ranks, it was sufficient to determine the wishes of the one of higher rank (*Manners* 1879).

5.45. the "feeling" is to be reciprocal. Figure 5.2 depicts "reciprocal feeling" (i.e., back and forth feeling) between a square and a pentagon. First, the square remains motionless while the pentagon "feels" him by situating himself so that he abuts the square with one of the square's vertices (say, P) near the midpoint of the abutting edge. The pentagon then pivots about P until he abuts the other edge that forms the angle at P. To complete the process, the pentagon remains motionless while the square pivots about one of the pentagon's vertices.

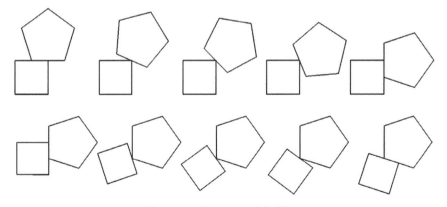

Figure 5.2. Reciprocal feeling.

purity of their native language – the formula is still further
curtailed by the use of "to feel" in a technical sense, mean-
50   ing, "to recommend-for-the-purposes-of-feeling-and-being-
felt"; and at this moment the "slang" of polite or fast society in the upper classes sanctions such a barbarism as "Mr.
Smith, permit me to feel you Mr. Jones."

Let not my Reader however suppose that "feeling" is with
55   us the tedious process that it would be with you, or that we
find it necessary to feel right round all the sides of every individual before we determine the class to which he belongs.
Long practice and training, begun in the schools and continued in the experience of daily life, enable us to discriminate
60   at once by the sense of touch, between the angles of an equal-
sided Triangle, Square, and Pentagon; and I need not say that
the brainless vertex of an acute-angled Isosceles is obvious
to the dullest touch. It is therefore not necessary, as a rule,
to do more than feel a single angle of an individual; and
65   this, once ascertained, tells us the class of the person whom
we are addressing, unless indeed he belongs to the higher
sections of the nobility. There the difficulty is much greater.
Even a Master of Arts in our University of Wentbridge has
been known to confuse a ten-sided with a twelve-sided Poly-
70   gon; and there is hardly a Doctor of Science in or out of that
famous University who could pretend to decide promptly
and unhesitatingly between a twenty-sided and a twenty-
four sided member of the Aristocracy.

Those of my readers who recall the extracts I gave above
75   from the Legislative code concerning Women, will readily
perceive that the process of introduction by contact requires
some care and discretion. Otherwise the angles might inflict
on the unwary Feeler irreparable injury. It is essential for
the safety of the Feeler that the Felt should stand perfectly
80   still. A start, a fidgety shifting of the position, yes, even a
violent sneeze, has been known before now to prove fatal
to the incautious, and to nip in the bud many a promising
friendship. Especially is this true among the lower classes
of the Triangles. With them, the eye is situated so far from
85   their vertex that they can scarcely take cognizance of what
goes on at that extremity of their frame. They are moreover

5.48. indifferent to the purity of their native language. For Abbott as a teacher of the English language, see Appendix B1, 1871.

5.51. "slang." The Square voices Abbott's contempt for the use of slang, which sometimes covers an "ignorance of the words of polite diction; but more often it is not so much ignorance as laziness that is the cause. Slang is intended to save the necessity of thinking, and it answers the purpose" (Abbott and Seeley 1871, 105).

5.52. barbarism. A use of a word or expression that violates the classical standards of a language; from the Greek word for foreigner (Fowler and Gowers 1965, 48–49).

5.56. right round. All the way around.

5.68. University of Wentbridge. Although there is an English village called Wentbridge, the University of Wentbridge represents the University of Cambridge. Abbott was a student at St. John's College, Cambridge, from 1857 until 1861. Jonathan Smith suggests that Abbott has turned "Came-bridge" to "Went-bridge" to indicate that his alma mater is headed in the wrong direction (Smith 1994, 265). In a letter to his fellow Johnian, Alfred Marshall, Abbott complains about the narrowly linguistic and literary method of studying classics that was employed at Cambridge when he was a student: "I can never look back without regret at 3.5 years spent in the study of little else but mere words, apart from the subject matter, of classical authors" (Abbott 1872).

5.69. to confuse a ten-sided. The measure of each interior angle of a regular ten-sided polygon is $144°$; for a twelve-sided polygon the measure is $150°$. For twenty-sided and twenty-four-sided polygons, the measures are $162°$ and $165°$, respectively.

5.85. the eye . . . so far from their vertex. It seems that the eye of an Isosceles is located at one of its base angles.

of a rough coarse nature, not sensitive to the delicate touch of the highly organized Polygon. What wonder then if an involuntary toss of the head has ere now deprived the State
90   of a valuable life!

I have heard that my excellent Grandfather – one of the least irregular of his unhappy Isosceles class, who indeed obtained, shortly before his decease, four out of seven votes from the Sanitary and Social Board for passing him into
95   the class of the Equal-sided – often deplored, with a tear in his venerable eye, a miscarriage of this kind, which had occurred to his great-great-great-Grandfather, a respectable Working Man with an angle or brain of 59° 30′. According to his account, my unfortunate Ancestor, being afflicted with
100   rheumatism, and in the act of being felt by a Polygon, by one sudden start accidentally transfixed the Great Man through the diagonal; and thereby, partly in consequence of his long imprisonment and degradation, and partly because of the moral shock which pervaded the whole of my Ancestor's
105   relations, threw back our family a degree and a half in their ascent towards better things. The result was that in the next generation the family brain was registered at only 58°, and not till the lapse of five generations was the lost ground recovered, the full 60° attained, and the Ascent from the
110   Isosceles finally achieved. And all this series of calamities from one little accident in the process of Feeling.

At this point I think I hear some of my better educated readers exclaim, "How could you in Flatland know anything about angles and degrees, or minutes? We can *see* an angle,
115   because we, in the region of Space, can see two straight lines inclined to one another; but you, who can see nothing but one straight line at a time, or at all events only a number of bits of straight lines all in one straight line, – how can you ever discern any angle, and much less register angles of different
120   sizes?"

I answer that though we cannot *see* angles, we can *infer* them, and this with great precision. Our sense of touch, stimulated by necessity, and developed by long training,

5.91. excellent Grandfather. "Excellent" is an honorific adjective of respect. Abbott's paternal grandfather, Edward Abbott (1782–1853), was an oilman (one who makes or sells animal or vegetable oil) and warehouseman.

5.92. unhappy. Unfortunate.

5.95. deplored. From the context, it seems that the Square intends an obsolete meaning of deplore: to tell of or recount with sorrow.

5.96. a tear in his venerable eye. Perhaps an allusion to Alexander Pope's *Odyssey*:

> At this the father, with a father's fears
> (His venerable eyes bedimm'd with tears):

5.96. miscarriage. Accident; literally, a mis-carrying.

5.98. angle or brain. The original theory of phrenology, pioneered by the German physician Franz Joseph Gall, ascribed various mental functions to specific regions of the brain. From this hypothesis, Gall reasoned that the surface of the skull indicates an individual's intellect and personality. Phrenology was popularized in Britain by Gall's collaborator, Johann Gaspar Spurzheim, and later by George Combe, whose *The Constitution of Man Considered in Relation to External Objects* was one of the best-selling books of the nineteenth century.

5.102. through the diagonal. A diagonal of a polygon is a line segment joining two non-adjacent vertices. It is not clear what is meant by "the" diagonal.

5.107. next generation . . . only $58°$. Another example of the inheritance of acquired characteristics.

5.108. lapse. Passing.

enables us to distinguish angles far more accurately than
125   your sense of sight, when unaided by a rule or measure
of angles. Nor must I omit to explain that we have great
natural helps. It is with us a Law of Nature that the brain
of the Isosceles class shall begin at half a degree, or thirty
minutes, and shall increase (if it increases at all) by half a
130   degree in every generation; until the goal of 60° is reached,
when the condition of serfdom is quitted, and the freeman
enters the class of Regulars.

Consequently, Nature herself supplies us with an ascending scale or Alphabet of angles for half a degree up to 60°,
135   Specimens of which are placed in every Elementary School
throughout the land. Owing to occasional retrogressions, to
still more frequent moral and intellectual stagnation, and to
the extraordinary fecundity of the Criminal and Vagabond
Classes, there is always a vast superfluity of individuals of
140   the half degree and single degree class, and a fair abundance
of Specimens up to 10°. These are absolutely destitute of civic
rights; and a great number of them, not having even intelligence enough for the purposes of warfare, are devoted by
the States to the service of education. Fettered immovably so
145   as to remove all possibility of danger, they are placed in the
class rooms of our Infant Schools, and there they are utilized
by the Board of Education for the purpose of imparting to
the offspring of the Middle Classes that tact and intelligence
of which these wretched creatures themselves are utterly
150   devoid.

In some states the Specimens are occasionally fed and suffered to exist for several years; but in the more temperate
and better regulated regions, it is found in the long run more
advantageous for the educational interests of the young,
155   to dispense with food, and to renew the Specimens every
month, – which is about the average duration of the foodless
existence of the Criminal class. In the cheaper schools, what
is gained by the longer existence of the Specimen is lost,
partly in the expenditure for food, and partly in the dimin-
160   ished accuracy of the angles, which are impaired after a few
weeks of constant "feeling." Nor must we forget to add, in

5.126. measure of angles. Although Flatlanders can measure angles, they could not use a Spaceland protractor, which must be placed on the angle being measured.

5.131. quitted. Left.

5.134. ascending scale. Aristotle suggests that all animals be arranged in a *scala naturae* according to their degree of "perfection"; he proposes a hierarchy of all organisms based on "powers of the soul." Later, Plotinus systematized the ideas of Aristotle and Plato to construct a "Great Chain of Being," a hierarchy of all forms of life (Lovejoy 1960, 58–63).

5.136. owing to occasional retrogressions. Flatland's Lamarckian theory of evolution with its inherent tendency toward progressive development does not account for the presence of large numbers of beings of the lowest type. Lamarck explained this discrepancy by postulating that the lowest forms of life were continually coming into existence from inorganic matter. Abbott has solved the problem in Flatland with "retrogressions" and "stagnation." For a brief speculation on the "population biology" of Flatland, see Dewdney (2002).

5.139. Criminal and Vagabond Classes. The hostility toward persons who move from place to place without a fixed home has ancient roots. In English law, "vagabond" is always used in a pejorative sense: an idle, worthless person; a rogue.

5.148. tact. A play on the literal meaning of tact, the sense of touch. To have tact is to have just the right touch.

enumerating the advantages of the more expensive system, that it tends, though slightly yet perceptibly, to the diminution of the redundant Isosceles population – an object which
165   every statesman in Flatland constantly keeps in view. On the whole therefore – although I am not ignorant that, in many popularly elected School Boards, there is a reaction in favour of "the cheap system," as it is called – I am myself disposed to think that this is one of the many cases in which expense
170   is the truest economy.

But I must not allow questions of School Board politics to divert me from my subject. Enough has been said, I trust, to show that Recognition by Feeling is not so tedious or indecisive a process as might have been supposed; and it is obvi-
175   ously more trustworthy than Recognition by hearing. Still there remains, as has been pointed out above, the objection that this method is not without danger. For this reason many in the Middle and Lower classes, and all without exception in the Polygonal and Circular orders, prefer a third method, the
180   description of which shall be reserved for the next section.

§6
Of Recognition by Sight

I AM ABOUT TO APPEAR very inconsistent. In previous sections I have said that all figures in Flatland present the appearance of a straight line; and it was added or implied, that it is consequently impossible to distinguish by the visual
5   organ between individuals of different classes: yet now I am about to explain to my Spaceland Critics how we are able to recognize one another by the sense of sight.

If however the Reader will take the trouble to refer to the passage in which Recognition by Feeling is stated to be

5.167. popularly elected School Boards. According to the Education Act of 1870, which established the elements of a primary school system, Great Britain was divided into 2,500 school districts each with a popularly elected school board. For Abbott's role in the election of two women to the first London School Board, see Appendix B1, 1870.

5.168. "the cheap system." A possible allusion to a controversial system for supporting elementary education introduced in 1862 by Robert Lowe, the Vice-President (responsible for policy) of the Privy Council's Committee on Education. According to the "Revised Code" (notoriously known as "payment by results"), the amount of a State grant to a school depended upon not only attendance but also the number of students in the school who passed examinations by inspectors in reading, writing, and arithmetic. In practice, it meant that teachers' salaries were largely determined by their success in preparing students for an annual examination. "If it (education) is not cheap it shall be efficient; if it is not efficient it shall be cheap," was Lowe's characteristically sarcastic remark. From the beginning, the Code was both reviled and praised, but it endured for more than thirty years. Abbott believed that the Code had "a natural tendency to produce mechanical teachers and stupid pupils" (*The Times*, 29 November 1873, 6).

*Notes on Section 6**

*Notes start line 12.

10   universal, he will find this qualification – "among the lower classes." It is only among the higher classes and in our temperate climates that Sight Recognition is practised.

That this power exists in any regions and for any classes, is the result of Fog; which prevails during the greater part
15   of the year in all parts save the torrid zones. That which is with you in Spaceland an unmixed evil, blotting out the landscape, depressing the spirits, and enfeebling the health, is by us recognized as a blessing scarcely inferior to air itself, and as the Nurse of arts and Parent of sciences. But let me
20   explain my meaning, without further eulogies on this beneficent Element.

If Fog were non-existent, all lines would appear equally and indistinguishably clear; and this is actually the case in those unhappy countries in which the atmosphere is per-
25   fectly dry and transparent. But wherever there is a rich supply of Fog, objects that are at a distance, say of three feet, are appreciably dimmer than those at a distance of two feet eleven inches; and the result is that by careful and constant experimental observation of comparative dimness and clear-
30   ness, we are enabled to infer with great exactness the configuration of the object observed.

An instance will do more than a volume of generalities to make my meaning clear.

Suppose I see two individuals approaching whose rank I
35   wish to ascertain. They are, we will suppose, a Merchant and a Physician, or in other words, an Equilateral Triangle and a Pentagon: how am I to distinguish them?

It will be obvious, to every child in Spaceland who has touched the threshold of Geometrical Studies, that, if
40   I can bring my eye so that its glance may bisect an angle (A) of the approaching stranger, my view will lie as it were evenly between his two sides that are next to me (viz. CA and AB), so that I shall contemplate the two impartially, and both will appear of the same size.

Notes on Section 6.

6.12. Sight Recognition. The exaggerated emphasis on sight recognition hints at Flatlanders' superficial perception and limited understanding of everything about them.

6.14. Fog. Throughout much of the nineteenth century, London was wrapped in a dense fog caused by the burning of soft coal. Abbott opened the 1882 Prize Day ceremonies at City of London School by announcing, "We meet here (at the original site of CLS on Milk Street) today for the last time. Yes, it is positively the *last* time that we shall conduct His Honour the Lord Mayor through devious and subterraneous passages to emerge into an apology for daylight and to breathe a substitute for air." Although the hall had plenty of large windows, the gaslights had been turned on at 2 P.M. because Abbott could not see to read the names of the prizewinners (City of London School 1882).

London's smog may have been responsible for Abbott's chronic respiratory ailments. In 1906, he wrote a friend, "I was never expected to live as a boy of ten, and at school was a perpetual invalid. My wildest dreams of working did not extend beyond my 63rd year" (Abbott 1906).

6.15. save. With the exception of.

6.19. Nurse of arts. In *King Henry V* 5.2, Shakespeare refers to peace as the "dear nurse of arts."

6.23. appear equally and indistinguishably clear. Without fog or some other medium that scatters light, an object appears equally bright at all distances from the eye, except when the object is a point source as far as the eye is concerned – then its apparent brightness varies inversely as the square of the distance from the eye (Houstoun 1930, 321–322).

The most important visual cues that enable humans to perceive an object as three-dimensional depend on slight differences in the two retinal images of the object. If Flatlanders had two eyes, they could use analogues of these cues to perceive the two-dimensional nature of an object. Nevertheless, Abbott has given each Flatlander only one eye and chosen to emphasize that the primary visual cue that enables a stationary observer to discern a stationary figure's shape is that the prevalent fog scatters light and causes closer objects to appear brighter. The deliberate irony of this construction is that fog, which is commonly associated with obscured vision, is essential to sight recognition.

6.39. has touched the threshold of. Has begun.

6.40. glance. Elsewhere in *Flatland*, "glance" has its usual meaning, a quick or a casual look; however, here it means line of sight or perhaps a ray of light issuing from the eye.

45   Now in the case of (1) the Merchant, what shall I see? I shall
see a straight line DAE, in which
the middle point (A) will be very
bright because it is nearest to me;

50   but on either side the line will
shade away *rapidly into dimness*,
because the sides AC and AB

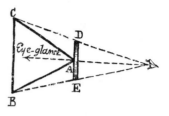

recede rapidly into the fog; and what appear to me as the Merchant's extremities, viz. D and E, will be *very dim indeed*.

55   On the other
hand in the case
of (2) the Physi-
cian, though I
shall here also

60   see a line (D'A'E')
with a bright
centre (A'), yet
it will shade
away *less rapidly*

65   into dimness,
because the sides

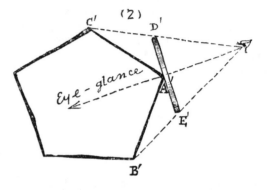

(A'C', A'B') *recede less rapidly into the fog*; and what appear to
me the Physician's extremities, viz. D' and E', will be *not so
dim* as the extremities of the Merchant.

70   The Reader will probably understand from these two
instances how – after a very long training supplemented
by constant experience – it is possible for the well-educated
classes among us to discriminate with fair accuracy between
the middle and lowest orders, by the sense of sight. If my

75   Spaceland Patrons have grasped this general conception, so
far as to conceive the possibility of it and not to reject my
account as altogether incredible – I shall have attained all I
can reasonably expect. Were I to attempt further details I
should only perplex. Yet for the sake of the young and inex-

80   perienced, who may perchance infer – from the two simple
instances I have given above, of the manner in which I should

6.47. I shall see a straight line DAE. Like the ancient Greeks, Flatlanders have arrived at a theory of vision without understanding either the nature of light or the anatomy of the eye or the brain. The Square's diagrams suggest that in the Flatlanders' theory of vision, there is no notion of a "retinal image" of an object of vision. In its place, there is a "straight line" (a line segment), which we shall call the "visual image" of the object, defined as follows: The "visual angle" of an object is the angle formed by the rays (half-lines) that have a common endpoint in the eye of the observer and pass through the extreme points of the object, that is, the points of the object that appear to be farthest right and left. For any convex object, there is one point of the figure that is closest to the observer's eye. If A denotes this point, then the visual image of the object is defined to be the line segment that contains A, is parallel to the line through the extreme points, and has its endpoints on the rays that form the visual angle.

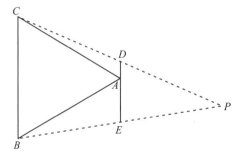

Figure 6.1. The visual angle of an equilateral triangle.

In Figure 6.1, where the object of vision is a triangle and the observer is located at P, the extreme points are B and C, and the rays from P passing through B and C (depicted by dotted lines) form the visual angle. The point of the triangle that is nearest to P is A, and the visual image of the triangle for a Flatlander at P is the segment DE.

Figure 6.2 illustrates an anomaly in this theory of vision. For a viewer at P, the octagon and the circumscribed circle have the same extreme points (B and C), hence the same angle of vision, yet the visual image of the circle (D'E') is smaller than visual image of the octagon (DE).

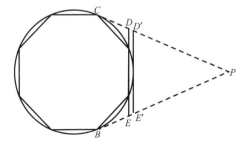

Figure 6.2. The visual angle of an octagon and a circumscribed circle.

recognize my Father and my Sons – that Recognition by sight
is an easy affair, it may be needful to point out that in actual
life most of the problems of Sight Recognition are far more
85   subtle and complex.

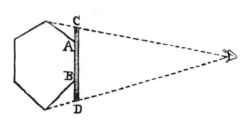

If for example,
when my Father, the
Triangle, approa-
ches me, he happens
90   to present his side
to me instead of his
angle, then, until
I have asked him
to rotate, or until I
95   have edged my eye
round him, I am for
the moment doubtful whether he may not be a Straight
Line, or, in other words, a Woman. Again, when I am
in the company of one of my two hexagonal Grandsons,
100   contemplating one of his sides (AB) full front, it will be
evident from the accompanying diagram that I shall see
one whole line (AB) in comparative brightness (shading off
hardly at all at the ends) and two smaller lines (CA and BD)
dim throughout and shading away into greater dimness
105   towards the extremities C and D.

But I must not give way to the temptation of enlarging on
these topics. The meanest mathematician in Spaceland will
readily believe me when I assert that the problems of life,
which present themselves to the well-educated – when they
110   are themselves in motion, rotating, advancing or retreating,
and at the same time attempting to discriminate by the sense
of sight between a number of Polygons of high rank moving
in different directions, as for example in a ball-room or conversazione – must be of a nature to task the angularity of the
115   most intellectual, and amply justify the rich endowments of
the Learned Professors of Geometry, both Static and Kinetic,
in the illustrious University of Wentbridge, where the

6.75. Spaceland Patrons. We, the readers of *Flatland*, are indeed the Square's patrons, for he lives under the dominion of our imaginations.

6.82. my Father. For Abbott's father, see Appendix B1, 1838.

6.95. edged my eye. To edge is to move edgeways; this is a play on the word "edge," which is a name for the side of a polygon.

6.98. Again. As another point or fact.

6.99. Grandsons. Abbott had no grandchildren; neither his son nor his daughter ever married. His closest living relatives are the descendants of his sisters, Elizabeth Parry and Alice Hart.

6.114. conversazione. An Italian word meaning an evening assembly for conversation or social recreation. In England, the word is used for a private gathering now known as an "At Home," a reception, where the host/hostess have announced that they will be at home during certain hours when visitors may come and go as they please.

6.114. task the angularity. Put a strain upon the intelligence. "Angularity" is a nice example of Abbott's inventive use of geometric language. It is used only once (line 16.5) with its usual meaning, "having sharp corners," and three times it is used to mean "the configuration of one's angles." A Flatland figure's intelligence is directly related to the size of his interior angles, and angularity is twice used figuratively to mean intelligence (here and line 9.52). Ironically, the isosceles triangles with severely acute vertex angles are intellectually obtuse or dull, whereas the polygons with obtuse angles are intellectually acute or sharp-witted.

6.115. rich endowments. An organization's endowment is the assets that are kept permanently and from which income is derived for its support. Abbott is referring to the huge endowments that the Universities of Oxford and Cambridge had accumulated by the nineteenth century. Critics of the universities noted that a large portion of the income from their endowments was devoted to nonacademic purposes, and very little was used to support scholarship (Universities Commission Report 1874).

6.116. Static Sight Recognition. Static Sight Recognition deals with bodies at rest. The observer takes a position relative to the figure being observed so that the point of the figure that is brightest (and so is nearest) is one of the figure's vertices, say, A; and the extreme points of the figure are equally bright (and so equally distant). The line segment from the eye to A then bisects the visual angle, and A is the midpoint of the visual image (see Note 6.54). From the varying brightness of the visual image, one can determine the slopes of the edges that form the angle at A and thereby determine this angle. For figures with even a modest number of sides, the art of sight recognition demands great precision. For example, using this method to determine that a polygon has sixteen sides requires estimating an angle with an error of less than 1%.

Science and Art of Sight Recognition are regularly taught to large classes of the *élite* of the States.

120   It is only a few of the scions of our noblest and wealthiest houses, who are able to give the time and money necessary for the thorough prosecution of this noble and valuable Art. Even to me, a Mathematician of no mean standing, and the Grandfather of two most hopeful and perfectly regular 125 Hexagons, to find myself in the midst of a crowd of rotating Polygons of the higher classes, is occasionally very perplexing. And of course to a common Tradesman, or Serf, such a sight is almost as unintelligible as it would be to you, my Reader, were you suddenly transported into our country.

130   In such a crowd you could see on all sides of you nothing but a Line, apparently straight, but of which the parts would vary irregularly and perpetually in brightness or dimness. Even if you had completed your third year in the Pentagonal and Hexagonal classes in the University, and were perfect in 135 the theory of the subject, you would still find that there was need of many years of experience, before you could move in a fashionable crowd without jostling against your betters, whom it is against etiquette to ask to "feel," and who, by their superior culture and breeding, know all about your 140 movements, while you know very little or nothing about theirs. In a word, to comport oneself with perfect propriety in Polygonal society, one ought to be a Polygon oneself. Such at least is the painful teaching of my experience.

It is astonishing how much the Art – or I may almost call it 145 instinct – of Sight Recognition is developed by the habitual practice of it and by the avoidance of the custom of "Feeling." Just as, with you, the deaf and dumb, if once allowed to gesticulate and to use the hand-alphabet, will never acquire the more difficult but far more valuable art of lip-speech 150 and lip-reading, so it is with us as regards "Seeing" and "Feeling." None who in early life resort to "Feeling" will ever learn "Seeing" in perfection.

6.116. Kinetic Sight Recognition. Kinetic Sight Recognition deals with bodies in motion. By tracking changes in the visual angle as a figure moves, a Flatland observer may be able to determine the nature and position of the figure. Examples of Kinetic Sight Recognition applied to rotating bodies are found in Section 15.

6.118. the Science and Art. Here "science" means learning acquired by study, and "art" means learning acquired by practice. Flatlanders studying the physics of sight are engaged in the Science of Sight Recognition. If they are using their past experience to position themselves and estimate a figure's angle, then they are practicing the Art of Sight Recognition.

6.121. only a few. . . are able to give the time. Aristotle recommends that mechanics, tradesmen, and farmers be excluded from the ideal state because they lack the leisure requisite both for the development of virtue and active participation in politics (*Politics*, 1329a).

6.123. Mathematician of no mean standing. Abbott's own mathematical background is sketched in Appendix B2, A "mathematical biography" of Edwin Abbott Abbott.

6.142. comport oneself with perfect propriety in Polygonal society. Abbott is mocking the system of etiquette in Victorian society, where the constantly changing standards of fashionably correct behavior made it impossible for an unworthy, ignorant outsiders to comport themselves properly (Davidoff 1973, 45).

6.150. will never acquire . . . lipspeech and lip-reading. Oralism is a method of teaching profoundly deaf people to communicate by the use of speech and lip-reading in preference to the use of sign language. At the International Congress of Teachers of the Deaf in Milan (1880), resolutions were introduced that called for the worldwide adoption of pure oralism and the rejection of sign languages in schools for the deaf. *The Times* enthusiastically reported the passing of these resolutions by large majorities, which included the majority of the British delegation (*The Times*, 28 September 1880, 9).

For this reason, among our Higher Classes, "Feeling" is discouraged or absolutely forbidden. From the cradle their children, instead of going to the Public Elementary schools (where the art of Feeling is taught,) are sent to higher Seminaries of an exclusive character; and at our illustrious University, to "feel" is regarded as a most serious fault, involving Rustication for the first offence, and Expulsion for the second.

But among the lower classes the art of Sight Recognition is regarded as an unattainable luxury. A common Tradesman cannot afford to let his son spend a third of his life in abstract studies. The children of the poor are therefore allowed to "feel" from their earliest years, and they gain thereby a precocity and an early vivacity which contrast at first most favourably with the inert, undeveloped, and listless behaviour of the half-instructed youths of the Polygonal class; but when the latter have at last completed their University course, and are prepared to put their theory into practice, the change that comes over them may almost be described as a new birth, and in every art, science, and social pursuit they rapidly overtake and distance their Triangular competitors.

Only a few of the Polygonal Class fail to pass the Final Test or Leaving Examination at the University. The condition of the unsuccessful minority is truly pitiable. Rejected from the higher class, they are also despised by the lower. They have neither the matured and systematically trained powers of the Polygonal Bachelors and Masters of Arts, nor yet the native precocity and mercurial versatility of the youthful Tradesman. The professions, the public services, are closed against them; and though in most States they are not actually debarred from marriage, yet they have the greatest difficulty in forming suitable alliances, as experience shows that the offspring of such unfortunate and ill-endowed parents is generally itself unfortunate, if not positively Irregular.

6.157. Seminaries of an exclusive character. In modern usage, a seminary is a school for training persons to be priests; figuratively, it is a place of early development. Flatland "seminaries" to which the higher classes send children "from the cradle" represent the English preparatory schools, that is, independent schools that prepared boys between eight and thirteen for the public schools.

6.159. Rustication. A form of punishment at Cambridge and Oxford whereby students were suspended from the university because of serious disciplinary infractions or failure of examinations. From a Latin stem meaning "to live in the country."

6.162. unattainable luxury. Plato insisted that education would remain the province of a wealthy elite (*Protagoras* 326c).

6.164. A common Tradesman . . . abstract studies. The City of London School was really three schools – an Elementary School preparing boys up to the age of thirteen for Middle School; a Middle School preparing a chosen few for Higher School while sending most of its pupils to commercial life at the age of sixteen; and a Higher School, teaching advanced classics and mathematics and preparing most of its pupils for some type of professional life and, in a fair number of cases, for the universities. Abbott strongly believed that all students, even those preparing for commercial business, should have a sound general education that would prepare them "to take a general interest in all knowledge, to use their own language with correctness and taste, and to understand something of the vastness and grandeur of the human as well as of the material world" (Abbott 1888, 381).

6.175. Leaving Examination. When Abbott attended Cambridge, students had two options for a B.A. degree: an Ordinary or an Honours degree. The examination for the Ordinary degree was "not so difficult but that any person of common abilities, and common preliminary training, with tolerable industry while at Cambridge, may reckon with certainty upon passing it" (*Student's Guide* 1866, 3). The examinations that Abbott took for the Honours degree are described in Appendix B2.

6.180. mercurial. Having the qualities ascribed to Mercury, the Roman divinity of commerce and gain: eloquence, ingenuity, aptitude for commerce.

It is from these specimens of the refuse of our Nobility that the great Tumults and Seditions of past ages have generally derived their leaders; and so great is the mischief thence
190   arising that an increasing minority of our more progressive Statesmen are of opinion that true mercy would dictate their entire suppression, by enacting that all who fail to pass the Final Examination of the University should be either imprisoned for life, or extinguished by a painless death.

195   But I find myself digressing into the subject of Irregularities, a matter of such vital interest that it demands a separate section.

§7
Of Irregular Figures

THROUGHOUT THE PREVIOUS PAGES I have been assuming – what perhaps should have been laid down at the beginning as a distinct and fundamental proposition – that every human being in Flatland is a Regular Figure, that is to say
5   of regular construction. By this I mean that a Woman must not only be a line, but a straight line; that an Artisan or Soldier must have two of his sides equal; that Tradesmen must have three sides equal; Lawyers (of which class I am a humble member), four sides equal, and, generally, that in every
10   Polygon, all the sides must be equal.

Notes on Section 7.

Title. Note that the title of Section 7 given in the Contents is "Concerning Irregular Figures."

7.4. Regular. In geometry, a polygon is said to be "regular" provided that all its sides are equal and all its angles are equal. Flatland figures, other than women and isosceles triangles, are "regular" provided that they are regular in the geometric sense. A woman is regular provided that she is truly straight, and an isosceles triangle is regular provided that he actually has two equal sides. Unlike geometric regularity, Flatland regularity is not an absolute property of figures; rather, it is a matter of judgment – at line 5.93, the Square tells us that his grandfather was certified as an equilateral triangle by a 4-3 vote of the Sanitary and Social Board.

7.8. three sides equal. The Square's "definition" of regularity is at best misleading. Any triangle with three equal sides has three equal angles, and conversely, any triangle with three equal angles is an equilateral triangle. For polygons with more than three sides, having all sides equal is not equivalent to having all angles equal. For example, an equal-sided quadrilateral (a rhombus) need not be equiangular, and an equiangular quadrilateral (a rectangle) need not be equal-sided. It is unclear whether Abbott has deliberately given an incorrect definition of regularity. In any case, Howard Candler's copy of *Flatland* (now held by Trinity College Library, Cambridge) contains a penciled addition "and four equal angles" above "four sides equal."

There are various conditions under which equilateral polygons are equiangular and vice versa. For example, an equilateral pentagon with three equal angles is equiangular (Euclid's *Elements*, Book XIII, Proposition 7). A general result of this kind is the following: A polygon is cyclic provided that all its vertices lie on a circle. Every regular polygon is cyclic, and every non-square rectangle is cyclic but not regular. Every equilateral, cyclic polygon is equiangular, and every equiangular, cyclic polygon with an odd number of sides is equilateral.

7.8. Lawyers. In England, lawyers are divided into barristers, who plead in the higher courts, and solicitors, who serve as legal advisers, do office work, and plead in the lower courts. Because barristers generally enjoyed high social status, it seems that the Square is a solicitor.

The size of the sides would of course depend upon the age of the individual. A Female at birth would be about an inch long, while a tall adult Woman might extend to a foot. As to the Males of every class, it may be roughly said that
15   the length of an adult's sides, when added together, is two feet or a little more. But the size of our sides is not under consideration. I am speaking of the *equality* of sides, and it does not need much reflection to see that the whole of the social life in Flatland rests upon the fundamental fact that
20   Nature wills all Figures to have their sides equal.

If our sides were unequal our angles might be unequal. Instead of its being sufficient to feel, or estimate by sight, a single angle in order to determine the form of an individual, it would be necessary to ascertain each angle by the
25   experiment of Feeling. But life would be too short for such a tedious groping. The whole science and art of Sight Recognition would at once perish; Feeling, so far as it is an art, would not long survive; intercourse would become perilous or impossible; there would be an end to all confidence, all
30   forethought; no one would be safe in making the most simple social arrangements; in a word, civilization would relapse into barbarism.

Am I going too fast to carry my Readers with me to these obvious conclusions? Surely a moment's reflection, and a sin-
35   gle instance from common life, must convince every one that our whole social system is based upon Regularity, or Equality of Angles. You meet, for example, two or three Tradesmen in the street, whom you recognize at once to be Tradesmen by a glance at their angles and rapidly bedimmed sides, and
40   you ask them to step into your house to lunch. This you do at present with perfect confidence, because everyone knows to an inch or two the area occupied by an adult Triangle: but imagine that your Tradesman drags behind his regular and respectable vertex, a parallelogram of twelve or thirteen
45   inches in diagonal: – what are you to do with such a monster sticking fast in your house door?

7.16. two feet. The first edition reads "three feet," which is certainly correct. "Two feet" is not consistent with the Square's earlier statement that the length of a Flatlander may be as great as twelve inches.

7.20. Nature wills. In the ancient Greek notion of nature, like the one in Flatland, there is not the modern concept of natural law as a closed system of causes and effects. Aristotle says that nature intends that the offspring of a good man would be good, but it cannot always accomplish this (*Politics*, 1255b). Nature in Flatland wills that all figures be regular, nevertheless, irregulars continue to be born.

7.21. might be unequal. This corrects the first edition, which read, "would be unequal."

7.28. intercourse. Dealings or communication between persons.

7.37. or Equality of Angles. Properly, Equality of Sides and Equality of Angles.

7.46. sticking fast. Unable to get any farther.

But I am insulting the intelligence of my Readers by accu-
mulating details which must be patent to everyone who
enjoys the advantages of a Residence in Spaceland. Obvi-
50   ously the measurements of a single angle would no longer be
sufficient under such portentous circumstances; one's whole
life would be taken up in feeling or surveying the perimeter
of one's acquaintances. Already the difficulties of avoiding a
collision in a crowd are enough to tax the sagacity of even a
55   well-educated Square; but if no one could calculate the Regularity of a single figure in the company, all would be chaos
and confusion, and the slightest panic would cause serious
injuries, or – if there happened to be any Women or Soldiers
present – perhaps considerable loss of life.

60   Expediency therefore concurs with Nature in stamping
the seal of its approval upon Regularity of conformation:
nor has the Law been backward in seconding their efforts.
"Irregularity of Figure" means with us the same as, or more
than, a combination of moral obliquity and criminality with
65   you, and is treated accordingly. There are not wanting, it is
true, some promulgators of paradoxes who maintain that
there is no necessary connection between geometrical and
moral Irregularity. "The Irregular," they say, "is from his
birth scouted by his own parents, derided by his brothers
70   and sisters, neglected by the domestics, scorned and suspected by society, and excluded from all posts of responsibility, trust, and useful activity. His every movement is jealously
watched by the police till he comes of age and presents himself for inspection; then he is either destroyed, if he is found
75   to exceed the fixed margin of deviation, or else immured in a
Government Office as a clerk of the seventh class; prevented
from marriage; forced to drudge at an uninteresting occupa-
tion for a miserable stipend; obliged to live and board at the
office, and to take even his vacation under close supervision;
80   what wonder that human nature, even in the best and purest,
is embittered and perverted by such surroundings!"

All this very plausible reasoning does not convince me, as
it has not convinced the wisest of our Statesmen, that our
ancestors erred in laying it down as an axiom of policy that

7.51. portentous. Threatening.

7.55. calculate. Count on; rely on.

7.63. Irregularity. The identification of irregularity with immorality in Flatland may have been suggested by Samuel Butler's *Erewhon* (1872). The citizens of Erewhon regard all diseases as crimes that require punishment; on the other hand, they regard all moral failings as diseases to be treated by "straighteners," whom they consult as we do physicians. The straighteners' remedies, which involve close confinement and in some cases physical torture, may be the sort of treatment used to cure irregularity in Flatland's Regular Hospitals.

Butler and Abbott were contemporaries at St. John's and knew one another. Butler reports that he twice had dinner at Abbott's home, but their views on theology were irreconcilable and they parted company (Jones 1968, 182–183).

7.64. obliquity. In geometry, a line that is inclined at an angle other than a right angle is said to be oblique; hence, obliquity means divergence from right conduct or thought.

7.66. promulgators of paradoxes. Those who make statements or proclaim doctrines that are contrary to received opinion or belief.

7.68. connection between geometrical and moral Irregularity. Physiognomy, the art of judging personal character from the physical features of the body, originated as a branch of ancient Greek medicine. Aristotle concludes the *Prior Analytics* with a chapter on the subject. From its beginnings, physiognomy has never lacked adherents. It was especially popular in Elizabethan England; Francis Bacon declared that when purged of superstition, it had "a solid ground in Nature, and a profitable use in life." The continued popularity of physiognomy in England, which lasted well into the nineteenth century, was due in part to Johann Caspar Lavater's *Essays on Physiognomy*.

7.69. scout. To mock at, deride.

7.70. domestics. House servants. Having servants in a Victorian household was not a mark of affluence; even the most modest middle-class family had at least one. The British census for 1891 lists two domestics living in Abbott's house, a housemaid aged twenty-two and a cook aged twenty-eight.

7.72. jealously. Zealously, distrustfully.

7.75. immured. Literally, enclosed within walls; confined.

7.82. plausible. Winning public approval with the implication of being superficially correct but not so in reality.

7.84. axiom of policy. In mathematics and logic, axioms are the starting points for logical deduction. The axioms are assumed to be true as the basis for logical argument, and the truth of the conclusions drawn from them depends on the truth of the axioms. By contrast, Euclid and the ancient Greek philosophers regarded an axiom as a self-evident truth, a statement that could be seen to be true without any need for proof. At line 8.64, it is clear that the Square intends "axiom" in the latter sense.

85   the toleration of Irregularity is incompatible with the safety
of the State. Doubtless, the life of an Irregular is hard; but the
interests of the Greater Number require that it shall be hard.
If a man with a triangular front and a polygonal back were
allowed to exist and to propagate a still more Irregular pos-
90   terity, what would become of the arts of life? Are the houses
and doors and churches in Flatland to be altered in order to
accommodate such monsters? Are our ticket-collectors to be
required to measure every man's perimeter before they allow
him to enter a theatre, or to take his place in a lecture room?
95   Is an Irregular to be exempted from the militia? And if not,
how is he to be prevented from carrying desolation into the
ranks of his comrades? Again, what irresistible temptations
to fraudulent impostures must needs beset such a creature!
How easy for him to enter a shop with his polygonal front
100   foremost, and to order goods to any extent from a confiding
tradesman! Let the advocates of a falsely called Philanthropy
plead as they may for the abrogation of the Irregular Penal
Laws, I for my part have never known an Irregular who was
not also what Nature evidently intended him to be – a hyp-
105   ocrite, a misanthropist, and, up to the limits of his power, a
perpetrator of all manner of mischief.

    Not that I should be disposed to recommend (at present)
the extreme measures adopted in some States, where an
infant whose angle deviates by half a degree from the cor-
110   rect angularity is summarily destroyed at birth. Some of our
highest and ablest men, men of real genius, have during their
earliest days laboured under deviations as great as, or even
greater than, forty-five minutes: and the loss of their precious
lives would have been an irreparable injury to the State. The
115   art of healing also has achieved some of its most glorious
triumphs in the compressions, extensions, trepannings, col-
ligations, and other surgical or diætetic operations by which
Irregularity has been partly or wholly cured. Advocating
therefore a *Via Media*, I would lay down no fixed or absolute
120   line of demarcation; but at the period when the frame is just
beginning to set, and when the Medical Board has reported
that recovery is improbable, I would suggest that the Irregular offspring be painlessly and mercifully consumed.

7.87. interests of the Greater Number. Abbott is satirizing the dominant political philosophy of the early Victorian Era, utilitarianism, which identifies the good with human happiness and maintains that actions are right provided they bring the greatest happiness to the greatest number. One objection to utilitarianism is that the prevention or elimination of suffering should take precedence over any alternative that only enhances the happiness of those who are already happy, and that is the point that Abbott is making. He raises this objection directly in *The Kernel and the Husk*, where he insists that no truly Christian "nation should tolerate and perpetuate the misery of a class in order that the whole nation may prosper" (Abbott 1886, 325).

7.90. posterity. A pun on "posterior."

7.95. exempted from. Excluded from participation.

7.98. must needs beset. Must possess. "Needs," which means "of necessity," merely emphasizes the sense of "must."

7.100. confiding. Trusting.

7.101. falsely called Philanthropy. Herbert Spencer believed that any intervention to cure social ills such as poverty would impede the progress that "inevitably" results from unchecked "social evolution." Spencer declared that "under the natural order of things society is constantly excreting its unhealthy, imbecile, slow, vacillating, faithless members." He inveighed against "spurious philanthropists" who, by stopping "the purifying process" to prevent present misery, cause even greater misery for future generations (Spencer 1851, 354–355). [continued]

7.108. extreme measures. A play on two meanings of extreme measures: literally, measurements taken with great care and figuratively, actions that are severe or violent in an exceedingly great degree.

7.110. destroyed at birth. Both Plato and Aristotle sanctioned infanticide by exposure of defective or deformed children, and the practice was far from rare in classical Greece. In Plato's ideal state, "the offspring of the inferior, and any of those of the other sort who are born defective, they will properly dispose of in secret" (*Republic*, 460c). Aristotle urges, "As to the exposure and rearing of children, let there be a law that no deformed child shall live" (*Politics*, 1335b).

7.116. trepannings. A trepan (from the Greek word for "borer") is a surgical instrument in the form of a crown-saw for cutting out small pieces of bone, especially from the skull. Hence, a trepanning is a removal of a circular area of bone; in Flatland, it is a removal of a small segment of the frame.

7.117. colligation. A binding together or connection.

7.117. diætetic. Of or pertaining to the diet; from the Latin *diæteticus*. The ordinary spelling is "dietetic."

7.119. *Via Media*. A highly ironic use of the Latin phrase meaning "a middle way."

7.123. Irregular offspring be . . . consumed. A play on two meanings of "consumed": "done away with" and "eaten." The Square's suggested means of disposal of incurably irregular children is an allusion to Jonathan Swift's satirical masterpiece, "A Modest Proposal" (1729). Swift proposes that the Irish solve the problem of widespread hunger by fattening up infant children and selling them at one year of age to the rich as food.

§8
Of the Ancient Practice
of Painting

If my Readers have followed me with any attention up to this point, they will not be surprised to hear that life is somewhat dull in Flatland. I do not, of course, mean that there are not battles, conspiracies, tumults, factions, and all
5   those other phenomena which are supposed to make History interesting; nor would I deny that the strange mixture of the problems of life and the problems of Mathematics, continually inducing conjecture and giving the opportunity of immediate verification, imparts to our existence a zest which
10   you in Spaceland can hardly comprehend. I speak now from the æsthetic and artistic point of view when I say that life with us is dull; æsthetically and artistically, very dull indeed.

How can it be otherwise, when all one's prospect, all one's landscapes, historical pieces, portraits, flowers, still life, are
15   nothing but a single line, with no varieties except degrees of brightness and obscurity?

It was not always thus. Colour, if Tradition speaks the truth, once for the space of half a dozen centuries or more, threw a transient splendour over the lives of our ancestors
20   in the remotest ages. Some private individual – a Pentagon whose name is variously reported – having casually discovered the constituents of the simpler colours and a rudimentary method of painting, is said to have begun decorating first his house, then his slaves, then his Father, his Sons, and
25   Grandsons, lastly himself. The convenience as well as the beauty of the results commended themselves to all. Wherever Chromatistes, – for by that name the most trustworthy

Notes on Section 8.

8.4. faction. Turbulent party strife or intrigue.

8.13. prospect. The visible scene or landscape.

8.14. landscapes, historical pieces ... still life. These are all works of art; in particular, a historical piece is a work of art depicting a historical event.

8.15. single line. A Flatland painting is a line segment on which the artist uses varying intensity to create the impression of (the perimeter of) a two-dimensional object. For a mathematical description of how such a picture is created in linear perspective, see Schlatter (2006).

8.19. transient splendour. The first edition reads "transient charm."

8.27. Chromatistes. The Greek word for color is *chroma*, and *istes* corresponds to the agent-forming English suffix -*ist*; hence, Chromatistes means colorist.

authorities concur in calling him, – turned his variegated
frame, there he at once excited attention, and attracted
30   respect. No one now needed to "feel" him; no one mistook
his front for his back; all his movements were readily ascertained by his neighbours without the slightest strain on their
powers of calculation; no one jostled him, or failed to make
way for him; his voice was saved the labour of that exhaust-
35   ing utterance by which we colourless Squares and Pentagons
are often forced to proclaim our individuality when we move
amid a crowd of ignorant Isosceles.

The fashion spread like wildfire. Before a week was over,
every Square and Triangle in the district had copied the
40   example of Chromatistes, and only a few of the more conservative Pentagons still held out. A month or two found
even the Dodecagons infected with the innovation. A year
had not elapsed before the habit had spread to all but the
very highest of the Nobility. Needless to say, the custom
45   soon made its way from the district of Chromatistes to sur-
rounding regions; and within two generations no one in all
Flatland was colourless except the Women and the Priests.

Here Nature herself appeared to erect a barrier, and to
plead against extending the innovation to these two classes.
50   Many-sidedness was almost essential as a pretext for the
Innovators. "Distinction of sides is intended by Nature to
imply distinction of colours" – such was the sophism which
in those days flew from mouth to mouth, converting whole
towns at a time to the new culture. But manifestly to our
55   Priests and Women this adage did not apply. The latter
had only one side, and therefore – plurally and pedanti-
cally speaking – *no sides*. The former – if at least they would
assert their claim to be really and truly Circles, and not
mere high-class Polygons with an infinitely large number
60   of infinitesimally small sides – were in the habit of boasting
(what Women confessed and deplored) that they also had
no sides, being blessed with a perimeter of one line or, in
other words, a Circumference. Hence it came to pass that
these two Classes could see no force in the so-called axiom
65   about "Distinction of Sides implying Distinction of Colour";
and when all others had succumbed to the fascinations of
corporal decoration, the Priests and the Women alone still
remained pure from the pollution of paint.

8.28. variegated. Many-colored.

8.52. sophism. An argument correct in form or appearance but actually fallacious, especially one used to deceive.

8.60. an infinitely large number of infinitesimally small sides. An exceedingly large number of extremely small sides.

8.62. blessed with a perimeter of one line. The Square uses "line" for what is now called "curve," the trace of a moving point.

    Aristotle declares that the circle is primary among plane figures because it has a perimeter of one line: "Every plane figure must be either rectilinear or curvilinear. Now the rectilinear is bounded by more than one line, the curvilinear by one only. But since in any kind the one is naturally prior to the many and the simple to the complex, the circle will be the first of plane figures" (*On the Heavens*, 286b).

8.63. it came to pass. This phrase occurs often in the historical narratives of the Old Testament in the King James Bible. Abbott has used it as a humorous allusion to the biblical style.

8.64. force. The power to convince or persuade.

8.64. so-called axiom. Here "axiom" plainly means "self-evident truth." The Square refers to the statement, "Distinction of sides implies distinction of colours" as a "so-called axiom" because the Women and the Circles do not believe that it is true, much less self-evident.

8.68. pure from the pollution. Pollution (*miasma*) is one of the most elusive concepts in Greek thought. Broadly speaking, it is a "defilement, the impairment of a thing's form or integrity," which could be removed by purification (Parker 1983, 3).

Immoral, licentious, anarchical, unscientific – call them by
what names you will – yet, from an æsthetic point of view,
those ancient days of the Colour Revolt were the glorious
childhood of Art in Flatland – a childhood, alas, that never
ripened into manhood, nor even reached the blossom of
youth. To live was then in itself a delight, because living
implied seeing. Even at a small party, the company was a
pleasure to behold; the richly varied hues of the assembly in
a church or theatre are said to have more than once proved
too distracting for our greatest teachers and actors; but most
ravishing of all is said to have been the unspeakable magnif-
icence of a military review.

The sight of a line of battle of twenty thousand Isosceles
suddenly facing about, and exchanging the sombre black
of their bases for the orange and purple of the two sides
including their acute angle; the militia of the Equilateral Triangles tricoloured in red, white, and blue; the mauve, ultra-
marine, gamboge, and burnt umber of the Square artillerymen rapidly rotating near their vermilion guns; the dashing
and flashing of the five-coloured and six-coloured Pentagons
and Hexagons careering across the field in their offices of
surgeons, geometricians and aides-de-camp – all these may
well have been sufficient to render credible the famous story
how an illustrious Circle, overcome by the artistic beauty
of the forces under his command, threw aside his marshal's
bâton and his royal crown, exclaiming that he henceforth
exchanged them for the artist's pencil. How great and glorious the sensuous development of these days must have been
is in part indicated by the very language and vocabulary of
the period. The commonest utterances of the commonest citizens in the time of the Colour Revolt seem to have been
suffused with a richer tinge of word or thought; and to that
era we are even now indebted for our finest poetry and for
whatever rhythm still remains in the more scientific utterance of these modern days.

8.69. licentious. Unrestrained by law, decorum, or morality.

8.69. unscientific. So great is the prestige of "science" in Flatland that "unscientific" is a highly pejorative epithet. Rosemary Jann observes that the prevailing (male) scientism in Flatland is intended to parody what Frank Turner called "scientific naturalism," the doctrine that the empirical methods of science are the only legitimate means of acquiring knowledge of the universe (Jann 1985). Between 1850 and 1900, a group of scientific naturalists led by the biologist Thomas H. Huxley sought to expand the influence of scientific ideas with the ultimate goal of secularizing English society (Turner 1974, 16). Charles Kingsley's *The Water-Babies* (1863) satirizes scientific naturalism in the character of Professor Ptthmllnsprts, who is based in part on Huxley.

8.71. Colour Revolt. A "color revolt" is one of several points of similarity between *Flatland* and the film *Pleasantville* (1998), in which two present-day twins are transported back to a black-and-white, 1950s television program. As the twins assume the lives of the television brother and sister they have replaced, they introduce new elements to Pleasantville – sensuality, passion, and the possibility of being one's own self – that are represented by the spontaneous arrival of color.

8.82. facing about. Turning the face in the opposite direction.

8.85. the militia of Equilateral Triangles. A military force raised from the civilian population to supplement a regular army.

8.85. tricoloured red, white, and blue. An allusion to the French tricolor, the national flag of France adopted at the Revolution, consisting of equal vertical stripes of blue, white, and red.

8.85. mauve. This dye was originally called "aniline purple" by William H. Perkin, who was only eighteen when he discovered it in 1856. The dye was later named "mauve," from the French word for the mallow flower. Perkin and Abbott certainly knew one another; they were the same age and were students together at the City of London School for two years before Perkin left at age fifteen to enter the Royal College of Chemistry. Perkin's sons, William and Arthur, were students of Abbott in the 1870s.

8.86. gamboge. A bright yellow color.

8.89. careering. Moving at full speed.

8.90. geometricians. Abbott uses "geometrician" rather than "geometer" because he intends the word in the sense of the original Greek, one who measures the earth, that is, a land surveyor or mapmaker. For us, a map of Flatland would be a depiction of that world on a plane surface displaying the relative size and position of certain features. But no being in Flatland can read a two-dimensional document, and so Flatland geometricians must draw one-dimensional maps.

8.94. Marshal's bâton. The field marshal's baton is the sign of the highest military rank in many armies.

8.95. pencil. Originally, "pencil" meant an artist's paintbrush.

8.98. language and vocabulary of the period. Perhaps an allusion to the Elizabethan Era, about which Abbott said "the English language was in such perfection that it seemed impossible for the men and women of those days to write weakly" (Abbott 1877b, 1).

8.102. scientific. Rigorous, precise.

§9
Of the Universal Colour Bill

BUT MEANWHILE THE INTELLECTUAL ARTS were fast decaying.

The Art of Sight Recognition, being no longer needed, was no longer practised; and the studies of Geometry, Statics, Kinetics, and other kindred subjects, came soon to be considered superfluous, and fell into disrepute and neglect even at our University. The inferior Art of Feeling speedily experienced the same fate at our Elementary Schools. Then the Isosceles classes, asserting that the Specimens were no longer used nor needed, and refusing to pay the customary tribute from the Criminal classes to the service of Education, waxed daily more numerous and more insolent on the strength of their immunity from the old burden which had formerly exercised the twofold wholesome effect of at once taming their brutal nature and thinning their excessive numbers.

Year by year the Soldiers and Artisans began more vehemently to assert – and with increasing truth – that there was no great difference between them and the very highest class of Polygons, now that they were raised to an equality with the latter, and enabled to grapple with all the difficulties and solve all the problems of life, whether Statical or Kinetical, by the simple process of Colour Recognition. Not content with the natural neglect into which Sight Recognition was falling, they began boldly to demand the legal prohibition of all "monopolising and aristocratic Arts" and the consequent abolition of all endowments for the studies of Sight Recognition, Mathematics, and Feeling. Soon, they began to insist that inasmuch as Colour, which was a second Nature,

Notes on Section 9.

9.10. tribute. Periodical payment made as an acknowledgment of submission.

9.28. Colour... had destroyed the need of aristocratic distinctions. In human history, color has been the basis of prejudicial discrimination. There is substantial irony in the fact that during the Colour Revolt in Flatland, color was a means of social leveling.

9.28. a second Nature. Note that the Square says, "was a second Nature," not "was second nature," which would mean "was natural or instinctive."

had destroyed the need of aristocratic distinctions, the Law
30   should follow in the same path, and that henceforth all individuals and all classes should be recognized as absolutely
equal and entitled to equal rights.

Finding the higher Orders wavering and undecided, the
leaders of the Revolution advanced still further in their
35   requirements, and at last demanded that all classes alike,
the Priests and the Women not excepted, should do homage
to Colour by submitting to be painted. When it was objected
that Priests and Women had no sides, they retorted that
Nature and Expediency concurred in dictating that the front
40   half of every human being (that is to say, the half containing his eye and mouth) should be distinguishable from his
hinder half. They therefore brought before a general and
extraordinary Assembly of all the States of Flatland a Bill
proposing that in every Woman the half containing the eye
45   and mouth should be coloured red, and the other half green.
The Priests were to be painted in the same way, red being
applied to that semicircle in which the eye and mouth formed
the middle point; while the other or hinder semicircle was to
be coloured green.

50   There was no little cunning in this proposal, which indeed
emanated, not from any Isosceles – for no being so degraded
would have had angularity enough to appreciate, much less
to devise, such a model of state-craft – but from an Irregular
Circle who, instead of being destroyed in his childhood, was
55   reserved by a foolish indulgence to bring desolation on his
country and destruction on myriads of his followers.

On the one hand the proposition was calculated to bring
the Women in all classes over to the side of the Chromatic
Innovation. For by assigning to the Women the same two
60   colours as were assigned to the Priests, the Revolutionists
thereby ensured that, in certain positions, every Woman
would appear like a Priest, and be treated with corresponding respect and deference – a prospect that could not fail to
attract the Female Sex in a mass.

9.37. do homage to. In Feudal Law, to make formal and public acknowledgment of allegiance.

9.43. extraordinary Assembly. The names of the governing bodies of Flatland are similar to those that governed Athens, an Assembly and a High Council. The *ecclesia* was the general assembly of all adult male citizens of Athens. These assemblies were either ordinary (held at regular intervals) or extraordinary (convened upon a sudden emergency) (Smith 1878, 439–443).

9.45. red/green. One way of distinguishing between port (left) and starboard (right) at sea is by using the colors red and green – the port navigation light of a ship is always red and the starboard light is always green. In the dark, this color scheme is the way that one ship can tell what direction another is headed.

9.55. reserved. Set apart for some destiny.

9.55. indulgence. A play on the theological meaning, remission of punishment, reinforcing the idea that geometric irregularity is equivalent to moral irregularity.

9.56. myriad. From the Greek word *myrioi*, which means "ten thousand" when the accent is on the first syllable and "a very large number" when the accent is on the second syllable (Ifrah 2000, 221).

65   But by some of my Readers the possibility of the identical
appearance of Priests and Women, under the new Legislation, may not be recognized; if so, a word or two will make
it obvious.

Imagine a woman duly decorated, according to the new
70   Code; with the front half (*i.e.* the half containing eye and
mouth) red, and with the hinder half green. Look at her from
one side. Obviously you will see a straight line, *half red, half
green.*

Now   imagine   a
75   Priest, whose mouth is
at M, and whose front
semicircle   (AMB)   is
consequently   coloured
red,   while   his   hinder
80   semicircle   is   green;   so
that   the   diameter   AB

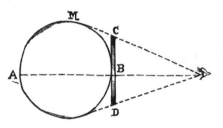

divides the green from the red. If you contemplate the Great
Man so as to have your eye in the same straight line as
his dividing diameter (AB), what you will see will be a
85   straight line (CBD), of which *one half* (CB) *will be red, and the
other* (BD) *green.* The whole line (CD) will be rather shorter
perhaps than that of a full-sized Woman, and will shade off
more rapidly towards its extremities; but the identity of the
colours would give you an immediate impression of identity
90   of Class, making you neglectful of other details. Bear in mind
the decay of Sight Recognition which threatened society at
the time of the Colour Revolt; add too the certainty that
Women would speedily learn to shade off their extremities
so as to imitate the Circles; it must then be surely obvious to
95   you, my dear Reader, that the Colour Bill placed us under a
great danger of confounding a Priest with a young Woman.

How attractive this prospect must have been to the Frail
Sex may readily be imagined. They anticipated with delight
the confusion that would ensue. At home they might hear
100   political and ecclesiastical secrets intended not for them
but for their husbands and brothers, and might even issue

Illustration. The illustration in the text is a perspective view of a three-dimensional eye; no two-dimensional being could have such an eye. Figure 9.1 depicts a possible Flatlander's two-dimensional eye, which is modeled on the cross-section of a human eye. Light passes through the transparent cornea, the pupil, and the lens to produce a one-dimensional image on the retina. See also Dewdney's conception of the Ardean eye in *The Planiverse* (Dewdney 1984b, 49).

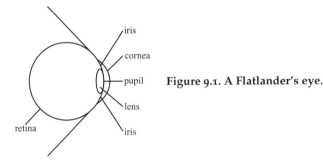

Figure 9.1. A Flatlander's eye.

9.90. of identity of Class. The first edition reads "of identity if not Class."

9.93. Women would speedily learn. This remark is the clearest of several signals to the reader that Flatland women are intelligent. To shade her extremities so as to present the appearance of a Circle, a woman would require a complete understanding of Sight Recognition, the highest form of learning at the University.

9.96. young Woman. In the Flatlander's theory of vision, the visual image of a young woman has the same length as the visual image of a circle. In Figure 9.2, the visual image of the circle is *AB*, and a young woman extending from *A* to *B* would present the same visual image apart from shading.

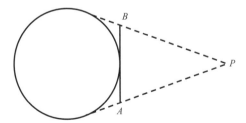

Figure 9.2. The visual angle of a circle and a young woman.

9.100. ecclesiastical. Pertaining to the government of Flatland (the *ecclesia*), viewed as consisting of the Priests.

commands in the name of a priestly Circle; out of doors the
striking combination of red and green, without addition of
any other colours, would be sure to lead the common people
105   into endless mistakes, and the Women would gain whatever
the Circles lost, in the deference of the passers by. As for the
scandal that would befall the Circular Class if the frivolous
and unseemly conduct of the Women were imputed to them,
and as to the consequent subversion of the Constitution, the
110   Female Sex could not be expected to give a thought to these
considerations. Even in the households of the Circles, the
Women were all in favour of the Universal Colour Bill.

The second object aimed at by the Bill was the gradual
demoralization of the Circles themselves. In the general intel-
115   lectual decay they still preserved their pristine clearness
and strength of understanding. From their earliest childhood, familiarized in their Circular households with the total
absence of Colour, the Nobles alone preserved the Sacred Art
of Sight Recognition, with all the advantages that result from
120   that admirable training of the intellect. Hence, up to the date
of the introduction of the Universal Colour Bill, the Circles
had not only held their own, but even increased their lead of
the other classes by abstinence from the popular fashion.

Now therefore the artful Irregular whom I described above
125   as the real author of this diabolical Bill, determined at one
blow to lower the status of the Hierarchy by forcing them to
submit to the pollution of Colour, and at the same time to
destroy their domestic opportunities of training in the Art of
Sight Recognition, so as to enfeeble their intellects by depriv-
130   ing them of their pure and colourless homes. Once subjected
to the chromatic taint, every parental and every childish Circle would demoralize each other. Only in discerning between
the Father and the Mother would the Circular infant find
problems for the exercise of its understanding – problems
135   too often likely to be corrupted by maternal impostures with
the result of shaking the child's faith in all logical conclusions. Thus by degrees the intellectual lustre of the Priestly
Order would wane, and the road would then lie open for
a total destruction of all Aristocratic Legislature and for the
140   subversion of our Privileged Classes.

9.124. Now therefore. Another example of "biblical diction." This phrase occurs often in the King James Bible.

9.124. artful Irregular. Perhaps an allusion to the "artful dodger," Jack Dawkins, in Dickens's *Oliver Twist*.

9.126. Hierarchy. The Circles or Priests.

§10
Of the Suppression of the Chromatic Sedition

The agitation for the Universal Colour Bill continued for three years; and up to the last moment of that period it seemed as though Anarchy were destined to triumph.

A whole army of Polygons, who turned out to fight as private soldiers, was utterly annihilated by a superior force of Isosceles Triangles – the Squares and Pentagons meanwhile remaining neutral. Worse than all, some of the ablest Circles fell a prey to conjugal fury. Infuriated by political animosity, the wives in many a noble household wearied their lords with prayers to give up their opposition to the Colour Bill; and some, finding their entreaties fruitless, fell on and slaughtered their innocent children and husbands, perishing themselves in the act of carnage. It is recorded that during that triennial agitation no less than twenty-three Circles perished in domestic discord.

Great indeed was the peril. It seemed as though the Priests had no choice between submission and extermination; when suddenly the course of events was completely changed by one of those picturesque incidents which Statesmen ought never to neglect, often to anticipate, and sometimes perhaps to originate, because of the absurdly disproportionate power with which they appeal to the sympathies of the populace.

It happened that an Isosceles of a low type, with a brain little if at all above four degrees – accidentally dabbling in the colours of some Tradesman whose shop he had plundered – painted himself, or caused himself to be painted (for the story varies) with the twelve colours of a Dodecagon. Going into the Market Place he accosted in a feigned voice

Notes on Section 10.

10.10. lords. This is the only place in *Flatland* that "lord" is not capitalized, and properly so – here it means husband.

10.14. triennial. Lasting for three years.

10.27. the twelve colours of a Dodecagon. The Square seems to be saying that all dodecagons were painted with the same colors – he speaks of *the* colors of the dodecagon. The first edition reads "Dodecahedron" instead of "Dodecagon." A dodecagon is a twelve-sided plane figure; a dodecahedron is a three-dimensional solid with twelve faces.

a maiden, the orphan daughter of a noble Polygon, whose
30   affection in former days he had sought in vain; and by a
series of deceptions – aided, on the one side, by a string of
lucky accidents too long to relate, and, on the other, by an
almost inconceivable fatuity and neglect of ordinary precau-
tions on the part of the relations of the bride – he succeeded
35   in consummating the marriage. The unhappy girl committed suicide on discovering the fraud to which she had been
subjected.

When the news of this catastrophe spread from State to
State the minds of the Women were violently agitated. Sym-
40   pathy with the miserable victim and anticipations of similar
deceptions for themselves, their sisters, and their daughters,
made them now regard the Colour Bill in an entirely new
aspect. Not a few openly avowed themselves converted to
antagonism; the rest needed only a slight stimulus to make
45   a similar avowal. Seizing this favourable opportunity, the
Circles hastily convened an extraordinary Assembly of the
States; and besides the usual guard of Convicts, they secured
the attendance of a large number of reactionary Women.

Amidst an unprecedented concourse, the Chief Circle of
50   those days – by name Pantocyclus – arose to find himself
hissed and hooted by a hundred and twenty thousand
Isosceles. But he secured silence by declaring that henceforth
the Circles would enter on a policy of Concession; yielding to
the wishes of the majority, they would accept the Colour Bill.
55   The uproar being at once converted to applause, he invited
Chromatistes, the leader of the Sedition, into the centre of the
hall, to receive in the name of his followers the submission
of the Hierarchy. Then followed a speech, a masterpiece of
rhetoric, which occupied nearly a day in the delivery, and to
60   which no summary can do justice.

With a grave appearance of impartiality he declared that,
as they were now finally committing themselves to Reform
or Innovation, it was desirable that they should take one last
view of the perimeter of the whole subject, its defects as well
65   as its advantages. Gradually introducing the mention of the
dangers to the Tradesmen, the Professional Classes and the

10.48. reactionary. Ultraconservative. One argument made by Victorian men and women against granting women the right to vote was that most of them were Conservatives and that their enfranchisement would consequently have a reactionary influence on politics (Fawcett 1870, 630).

10.49. concourse. A crowd, throng.

10.50. Pantocyclus. From the ancient Greek *panto*, meaning "all," and the Latin *cyclus*, "circle."

10.59. rhetoric. The art of using language to persuade or influence others.

Gentlemen, he silenced the rising murmurs of the Isosceles by reminding them that, in spite of all these defects, he was willing to accept the Bill if it was approved by the majority.
70 But it was manifest that all, except the Isosceles, were moved by his words and were either neutral or averse to the Bill.

Turning now to the Workmen he asserted that their interests must not be neglected, and that, if they intended to accept the Colour Bill, they ought at least to do so with full
75 view of the consequences. Many of them, he said, were on the point of being admitted to the class of the Regular Triangles; others anticipated for their children a distinction they could not hope for themselves. That honourable ambition would now have to be sacrificed. With the universal adoption of
80 Colour, all distinctions would cease; Regularity would be confused with Irregularity; development would give place to retrogression; the Workman would in a few generations be degraded to the level of the Military, or even the Convict Class; political power would be in the hands of the great-
85 est number, that is to say the Criminal Classes, who were already more numerous than the Workmen, and would soon out-number all the other Classes put together when the usual Compensative Laws of Nature were violated.

A subdued murmur of assent ran through the ranks of
90 the Artisans, and Chromatistes, in alarm, attempted to step forward and address them. But he found himself encompassed with guards and forced to remain silent while the Chief Circle in a few impassioned words made a final appeal to the Women, exclaiming that, if the Colour Bill passed,
95 no marriage would henceforth be safe, no woman's honour secure; fraud, deception, hypocrisy would pervade every household; domestic bliss would share the fate of the Constitution and pass to speedy perdition: "Sooner than this," he cried, "Come death."

100 At these words, which were the preconcerted signal for action, the Isosceles Convicts fell on and transfixed the wretched Chromatistes; the Regular Classes, opening their ranks, made way for a band of Women who, under

10.85. political power would be in the hands of the greatest number. Many Victorians were no more democratic than Plato, who expresses disdain for Athenian democracy throughout the *Republic*. One of the least egalitarian Victorians was Robert Lowe, an "illiberal Liberal" member of the House of Commons (see Note 5.168). Lowe believed that democracy would transfer political power to the ignorant, and so intelligent government would impossible. His eloquent speeches in opposition to the Reform Act of 1866 set out a comprehensive case against democracy. The opposition of Lowe and others led to the defeat of the Act and caused the Liberal government's downfall. The following year, a new government led by Benjamin Disraeli succeeded in passing the Reform Act of 1867, which nearly doubled the number of eligible voters by enfranchising all male householders.

10.98. perdition. Utter destruction, complete ruin.

10.100. preconcerted. Agreed upon beforehand.

10.102. transfixed . . . Chromatistes. The historical figures Chromatistes and Pantocyclus are the only "named" characters in *Flatland*, and the significance of these names reaches beyond their generic meanings, "colorist" and "all-circle." Although he is not a religious figure, Chromatistes is a messiah in the extended sense of the word, a zealous leader of a cause. It is surely no coincidence that his name includes the letters of "Christ," which is a translation of the Hebrew word for Messiah. Further, the assonance between "Pantocyclus" and "Pontius Pilate" suggests that Abbott intends the former, who ordered that Chromatistes be transfixed, as an allusion to the latter, who ordered that Christ be fixed to a cross.

direction of the Circles, moved, back foremost, invisibly and
105 unerringly upon the unconscious soldiers; the Artisans, imi-
tating the example of their betters, also opened their ranks.
Meantime bands of Convicts occupied every entrance with
an impenetrable phalanx.

The battle, or rather carnage, was of short duration. Under
110 the skilful generalship of the Circles almost every Woman's
charge was fatal, and very many extracted their sting unin-
jured, ready for a second slaughter. But no second blow was
needed; the rabble of the Isosceles did the rest of the business for themselves. Surprised, leader-less, attacked in front
115 by invisible foes, and finding egress cut off by the Convicts
behind them, they at once – after their manner – lost all presence of mind, and raised the cry of "treachery." This sealed
their fate. Every Isosceles now saw and felt a foe in every
other. In half an hour not one of that vast multitude was
120 living; and the fragments of seven score thousand of the
Criminal Class slain by one another's angles attested the
triumph of Order.

The Circles delayed not to push their victory to the uttermost. The Working Men they spared but decimated. The
125 Militia of the Equilaterals was at once called out; and every
Triangle suspected of Irregularity on reasonable grounds,
was destroyed by Court Martial, without the formality of
exact measurement by the Social Board. The homes of the
Military and Artisan classes were inspected in a course of
130 visitations extending through upwards of a year; and during
that period every town, village, and hamlet was systematically purged of that excess of the lower orders which had
been brought about by the neglect to pay the Tribute of Crim-
inals to the Schools and University, and by the violation of
135 the other natural Laws of the Constitution of Flatland. Thus
the balance of classes was again restored.

Needless to say that henceforth the use of Colour was
abolished, and its possession prohibited. Even the utterance
of any word denoting Colour, except by the Circles or by
140 qualified scientific teachers, was punished by a severe

10.105. unconscious. Unaware that they were now alone in favoring the Bill.

10.108. phalanx. A line or array of battle; specifically, a body of heavy-armed infantry drawn up in close order with shields joined and long spears overlapping.

10.124. decimate. Abbott is using the word in the literal sense, "to kill, destroy, or remove one in every ten," rather than the familiar meaning, "to destroy or remove a large proportion."

10.133. neglect. Failure.

penalty. Only at our University in some of the very highest and most esoteric classes – which I myself have never been privileged to attend – it is understood that the sparing use of Colour is still sanctioned for the purpose of illustrating some
145   of the deeper problems of mathematics. But of this I can only speak from hearsay.

Elsewhere in Flatland, Colour is now non-existent. The art of making it is known to only one living person, the Chief Circle for the time being; and by him it is handed down on
150   his death-bed to none but his Successor. One manufactory alone produces it; and, lest the secret should be betrayed, the Workmen are annually consumed, and fresh ones introduced. So great is the terror with which even now our Aristocracy looks back to the far distant days of the agitation for
155   the Universal Colour Bill.

§11
Concerning our Priests

IT IS HIGH TIME THAT I should pass from these brief and discursive notes about things in Flatland to the central event of this book, my initiation into the mysteries of Space. *That* is my subject; all that has gone before is merely preface.

5   For this reason I must omit many matters of which the explanation would not, I flatter myself, be without interest for my Readers: as for example, our method of propelling and stopping ourselves, although destitute of feet; the means by which we give fixity to structures of wood, stone, or brick,
10   although of course we have no hands, nor can we lay foundations as you can, nor avail ourselves of the lateral pressure of the earth; the manner in which the rain originates in the intervals between our various zones, so that the northern

10.144. illustrating. In 1847, the firm of William Pickering published one of the most beautiful books of the century, an edition of Euclid's *Elements* produced by the amateur mathematician Oliver Byrne. What distinguishes Byrne's edition is his "demonstrations" of Euclid's theorems using "coloured symbols, signs, and diagrams" rather than the usual proofs, which use "words, letters, and black or uncoloured diagrams." According to Byrne, "Such is the expedition of this enticing mode of communicating knowledge, that the Elements of Euclid can be acquired in less than one third the time usually employed, and the retention by the memory is much more permanent; these facts have been ascertained by numerous experiments made by the inventor, and several others who have adopted his plans" (Byrne 1847, ix). The entire book is available online at the Digital Mathematics Archive.

10.150. manufactory. A workshop where things are handmade.

Notes on Section 11.

11.3. initiation into the mysteries. Abbott's figurative use of the ancient Greek mysteries is discussed in Note 20.109.

regions do not intercept the moisture from falling on the
15   southern; the nature of our hills and mines, our trees and
vegetables, our seasons and harvests; our Alphabet, suited
to our linear tablets; our eyes, adapted to our linear sides;
these and a hundred other details of our physical existence
I must pass over, nor do I mention them now except to indi-
20   cate to my readers that their omission proceeds, not from
forgetfulness on the part of the Author, but from his regard
for the time of the Reader.

Yet before I proceed to my legitimate subject some few
final remarks will no doubt be expected by my Readers upon
25   those pillars and mainstays of the Constitution of Flatland,
the controllers of our conduct and shapers of our destiny, the
objects of universal homage and almost of adoration: need I
say that I mean our Circles or Priests?

When I call them Priests, let me not be understood as mean-
30   ing no more than the term denotes with you. With us, our
Priests are Administrators of all Business, Art, and Science;
Directors of Trade, Commerce, Generalship, Architecture,
Engineering, Education, Statesmanship, Legislature, Morality, Theology; doing nothing themselves, they are the Causes
35   of everything, worth doing, that is done by others.

Although popularly every one called a Circle is deemed
a Circle, yet among the better educated Classes it is known
that no Circle is really a Circle, but only a Polygon with a very
large number of very small sides. As the number of the sides
40   increases, a Polygon approximates to a Circle; and, when the
number is very great indeed, say for example three or four
hundred, it is extremely difficult for the most delicate touch
to feel any polygonal angles. Let me say rather, it *would* be
difficult: for, as I have shown above, Recognition by Feeling
45   is unknown among the highest society, and to *feel* a Circle
would be considered a most audacious insult. This habit of
abstention from Feeling in the best society enables a Circle
the more easily to sustain the veil of mystery in which, from
his earliest years, he is wont to enwrap the exact nature of
50   his Perimeter or Circumference. Three feet being the average

11.15. mines. A Flatland "mineshaft" does not go "down" but runs "laterally." The roof of the shaft could be supported by a series of door/walls. A Flatland miner could walk through the mine opening one door at a time while the roof was supported by the unopened doors (Dewdney 1984b, 72).

11.16. Alphabet. The first edition reads "Alphabet, and method of writing, adapted to our linear tablets" instead of "Alphabet, suited to our linear tablets; our eyes, adapted to our linear sides."

11.18. details of our physical existence. It is not the details of physical existence but rather the "human" details of *Flatland* that give it verisimilitude. Indeed, *Flatland* seems more "real" than stories in which the physical details of two-dimensional space are more fully developed.

C. Howard Hinton, who is best known as a popularizer of the fourth dimension, attempted to give a consistent account of the "physical" details of a two-dimensional world. In his essay, "A plane world," he says, "I should have wished to be able to refer the reader altogether to that ingenious work, 'Flatland.'" But, he continues, Abbott has used the conditions of life on a plane "as a setting wherein to place his satire and his lessons." He characterizes his own approach as primarily interested in the physical facts (Hinton 1886, 129). [continued]

11.31. our Priests. Like the prisoners in Plato's cave, Flatlanders are fettered by their "dimensional prejudice" – the conviction that the world of their sense perception is the only possible world. The shadowy reality that each figure "knows" is obscured by this universally held prejudice. The shared "reality" of Flatland society is itself indirectly determined by the Priests. They correspond to Plato's "image-makers," those who interpret reality for others: writers, artists, scientists, educators, businessmen, statesmen, legislators, as well as philosophers and religious leaders (*Republic*, 514ab).

11.35. the Causes of everything, worth doing, that is done by others. The social reformer and author, Beatrice Potter Webb, recalls that as she grew older she became aware that she "belonged to a class of persons who habitually gave orders, but who seldom, if ever, executed the orders of other people." The essential qualification for membership in London Society, she says, was "the possession of some form of power over other people" (Webb 1926, 42, 49).

11.38. no Circle is really a Circle. In the *Philebus*, Plato contrasts the Form of the circle ("the divine circle") with the material circle used in construction ("the human circle").

11.40. As the number of the sides . . . Circle. This corrects a mistake in the first edition, which reads, "In proportion to the number of the sides the Polygon approximates to a Circle."

11.40. approximates to a Circle. Ancient mathematicians recognized that the ratio of the circumference of a circle to its diameter is constant, and approximations to this constant (π) were given by the Babylonians, Egyptians, and Chinese. (The symbol π was not used until 1706.) In *Measurement of the Circle*, Archimedes describes a method that is a fundamental improvement on earlier methods. He inscribes in and circumscribes about a circle a sequence of regular polygons with diminishing sides. [continued]

Perimeter it follows that, in a polygon of three hundred sides, each side will be no more than the hundredth part of a foot in length, or little more than the tenth part of an inch; and in a Polygon of six or seven hundred sides the sides are little larger than the diameter of a Spaceland pin-head. It is always assumed, by courtesy, that the Chief Circle for the time being has ten thousand sides.

The ascent of the posterity of the Circles in the social scale is not restricted, as it is among the lower Regular classes, by the Law of Nature which limits the increase of sides to one in each generation. If it were so, the number of sides in a Circle would be a mere question of pedigree and arithmetic; and the four hundred and ninety-seventh descendant of an Equilateral Triangle would necessarily be a Polygon with five hundred sides. But this is not the case. Nature's Law prescribes two antagonistic decrees affecting Circular propagation; first, that as the race climbs higher in the scale of development, so development shall proceed at an accelerated pace; second, that in the same proportion, the race shall become less fertile. Consequently in the home of a Polygon of four or five hundred sides it is rare to find a son; more than one is never seen. On the other hand the son of a five-hundred-sided Polygon has been known to possess five hundred and fifty, or even six hundred sides.

Art also steps in to help the process of the higher Evolution. Our physicians have discovered that the small and tender sides of an infant Polygon of the higher class can be fractured, and his whole frame re-set, with such exactness that a Polygon of two or three hundred sides sometimes – by no means always, for the process is attended with serious risk – but sometimes overleaps two or three hundred generations, and as it were doubles at a stroke, the number of his progenitors and the nobility of his descent.

Many a promising child is sacrificed in this way. Scarcely one out of ten survives. Yet so strong is the parental ambition among those Polygons who are, as it were, on the fringe of the Circular class, that it is very rare to find a Nobleman,

11.56. by courtesy. An idiomatic expression meaning "by common good will or allowance," as distinguished from inherent or legal right.

11.78. fractured, and his whole frame re-set. In a letter to his brother dated 14 May 1887, Samuel Barnett (the founder of Toynbee Hall) quotes Abbott's description of his friend Montagu Butler, the Master of Trinity: "He is a polygon who has been made a circle. He has natural strong edges which have been compressed" (Barnett 1919, 33).

11.83. descent. Descendants.

of that position in society, who has neglected to place his first-born son in the Circular Neo-Therapeutic Gymnasium
90   before he has attained the age of a month.

One year determines success or failure. At the end of that time the child has, in all probability, added one more to the tombstones that crowd the Neo-Therapeutic Cemetery; but on rare occasions a glad procession bears back the little one
95   to his exultant parents, no longer a Polygon, but a Circle, at least by courtesy: and a single instance of so blessed a result induces multitudes of Polygonal parents to submit to similar domestic sacrifices, which have a dissimilar issue.

§12
Of the Doctrine of our Priests

As to the doctrine of the Circles it may briefly be summed up in a single maxim, "Attend to your Configuration." Whether political, ecclesiastical, or moral, all their teaching has for its object the improvement of individual and
5   collective Configuration – with special reference of course to the Configuration of the Circles, to which all other objects are subordinated.

It is the merit of the Circles that they have effectually suppressed those ancient heresies which led men to waste
10   energy and sympathy in the vain belief that conduct depends

11.90. Neo-Therapeutic Gymnasium before . . . the age of one month. In Greek cities, *gymnasia* originated as places of exercise but later they became more intellectual centers, and some like Plato's Academy and Aristotle's Lyceum were precursors of modern-day universities. In Germany and other Continental countries, a gymnasium is a school designed to prepare students for the universities.

11.93. added one more to the tombstones. Abbott is using the death of a child in Flatland's Circular Neo-Therapeutic Gymnasium as a figure for the physical and psychological toll exacted from public school students by those who would instill "manliness."

Flatland gymnasia most likely represent the great English "public schools" (actually, private boarding schools), which furnished Oxford and Cambridge with most of their students. The public schools were originally intended to provide a free education for local boys; however, by the beginning of the nineteenth century, they had largely become boarding schools for the sons of the aristocracy. [continued]

11.98. issue. A play on two meanings of issue: offspring and outcome or consequence.

Notes on Section 12.

12.9. heresies. Doctrines at variance with those accepted as authoritative.

12.10. conduct depends. An allusion to the age-old "nature versus nurture" debate: Are inherited characteristics (nature) or environmental factors (nurture) the dominant influence on human behavior?

Like the English philosophers John Locke and John Stuart Mill, Abbott strongly believed in the efficacy of nurture over nature; however, by the time *Flatland* appeared this view had become an "ancient heresy." The leading advocate of the predominant view, the geneticist Francis Galton, insisted "There is no escape from the conclusion that nature prevails enormously over nurture when the differences of nurture do not exceed what is commonly to be found among persons of the same rank of society and in the same country" (Galton 1875). The doctrine of the Circles ("Configuration makes the man.") is a satirical representation of the "orthodox" position on nature versus nurture in the late nineteenth century.

In *Nature Via Nurture*, the science writer Matt Ridley makes a strong case that research into how genes are triggered by environmental events and how learning operates through gene expression has recast the terms of the debate. "It is no longer nature versus nurture but nature via nurture" (Ridley 2003).

upon will, effort, training, encouragement, praise, or anything else but Configuration. It was Pantocyclus – the illustrious Circle mentioned above, as the queller of the Colour
Revolt – who first convinced mankind that Configuration
15   makes the man; that if, for example, you are born an Isosceles
with two uneven sides, you will assuredly go wrong unless
you have them made even – for which purpose you must go
to the Isosceles Hospital; similarly, if you are a Triangle, or
Square, or even a Polygon, born with any Irregularity, you
20   must be taken to one of the Regular Hospitals to have your
disease cured; otherwise you will end your days in the State
Prison or by the angle of the State Executioner.

All faults or defects, from the slightest misconduct to the
most flagitious crime, Pantocyclus attributed to some devi-
25   ation from perfect Regularity in the bodily figure, caused
perhaps (if not congenital) by some collision in a crowd; by
neglect to take exercise, or by taking too much of it; or even by
a sudden change of temperature, resulting in a shrinkage or
expansion in some too susceptible part of the frame. There-
30   fore, concluded that illustrious Philosopher, neither good
conduct nor bad conduct is a fit subject, in any sober estima-
tion, for either praise or blame. For why should you praise,
for example, the integrity of a Square who faithfully defends
the interests of his client, when you ought in reality rather to
35   admire the exact precision of his right angles? Or again, why
blame a lying, thievish Isosceles when you ought rather to
deplore the incurable inequality of his sides?

Theoretically, this doctrine is unquestionable; but it has
practical drawbacks. In dealing with an Isosceles, if a rascal
40   pleads that he cannot help stealing because of his uneven-
ness, you reply that for that very reason, because he cannot
help being a nuisance to his neighbours, you, the Magistrate,
cannot help sentencing him to be consumed – and there's
an end of the matter. But in little domestic difficulties,

12.13. illustrious. Literally, "having luster or brilliance" – in this case, because of high birth or rank.

12.16. with two uneven sides. Having the two longest sides unequal.

12.17. made even. Made equal.

12.22. angle of the State Executioner. The Flatland equivalent of a guillotine.

12.24. flagitious. Extremely wicked, heinous, villainous.

12.35. his right angles. The first edition reads "his Rectangles," which was not a mistake – "rectangle" once meant a right angle.

45   where the penalty of consumption, or death, is out of the
question, this theory of Configuration sometimes comes in
awkwardly; and I must confess that occasionally when one
of my own Hexagonal Grandsons pleads as an excuse for his
disobedience that a sudden change of the temperature has
50   been too much for his Perimeter, and that I ought to lay the
blame not on him but on his Configuration, which can only
be strengthened by abundance of the choicest sweetmeats,
I neither see my way logically to reject, nor practically to
accept, his conclusions.

55   For my own part, I find it best to assume that a good
sound scolding or castigation has some latent and strength-
ening influence on my Grandson's Configuration; though I
own that I have no grounds for thinking so. At all events
I am not alone in my way of extricating myself from this
60   dilemma; for I find that many of the highest Circles, sitting as
Judges in Law courts, use praise and blame towards Regular
and Irregular Figures; and in their homes I know by experience that, when scolding their children, they speak about
"right" or "wrong" as vehemently and passionately as if they
65   believed that these names represented real existences, and
that a human Figure is really capable of choosing between
them.

Consistently carrying out their policy of making Configuration the leading idea in every mind, the Circles reverse the
70   nature of that Commandment which in Spaceland regulates
the relations between parents and children. With you, children are taught to honour their parents; with us – next to the
Circles, who are the chief object of universal homage – a man
is taught to honour his Grandson, if he has one; or, if not, his
75   Son. By "honour," however, is by no means meant "indulgence," but a reverent regard for their highest interests: and
the Circles teach that the duty of fathers is to subordinate
their own interests to those of posterity, thereby advancing
the welfare of the whole State as well as that of their own
80   immediate descendants.

12.52. sweetmeats. Sweet food, as sugared cakes or pastry; preserved or candied fruits, sugared nuts.

12.56. castigation. Corrective punishment or discipline.

The weak point in the system of the Circles – if a humble Square may venture to speak of anything Circular as containing any element of weakness – appears to me to be found in their relations with Women.

85 As it is of the utmost importance for Society that Irregular births should be discouraged, it follows that no Woman who has any Irregularities in her ancestry is a fit partner for one who desires that his posterity should rise by regular degrees in the social scale.

90 Now the Irregularity of a Male is a matter of measurement; but as all Women are straight, and therefore visibly Regular so to speak, one has to devise some other means of ascertaining what I may call their invisible Irregularity, that is to say their potential Irregularities as regards possible offspring. 95 This is effected by carefully-kept pedigrees, which are preserved and supervised by the State; and without a certified pedigree no Woman is allowed to marry.

Now it might have been supposed that a Circle – proud of his ancestry and regardful for a posterity which might 100 possibly issue hereafter in a Chief Circle – would be more careful than any other to choose a wife who had no blot on her escutcheon. But it is not so. The care in choosing a Regular wife appears to diminish as one rises in the social scale. Nothing would induce an aspiring Isosceles, who had hopes 105 of generating an Equilateral Son, to take a wife who reckoned a single Irregularity among her Ancestors; a Square or Pentagon, who is confident that his family is steadily on the rise, does not inquire above the five-hundredth generation; a Hexagon or Dodecagon is even more careless of the wife's 110 pedigree; but a Circle has been known deliberately to take a wife who has had an Irregular Great-Grandfather, and all because of some slight superiority of lustre, or because of the charms of a low voice – which, with us, even more than you, is thought "an excellent thing in Woman."

12.87. a fit partner. Proposals for improving the quality of the human race date from ancient times. For example, Socrates urges that selective breeding be used to ensure that the flock "be as perfect as possible" (*Republic*, 459e). Eugenics, the formal study of human improvement by the selection of desired heritable characteristics, began in the nineteenth century with the work of men like Francis Galton, who maintained, "[I]t would be quite practicable to produce a highly-gifted race of men by judicious marriages during several consecutive generations" (Galton 1869, 1).

12.97. certified pedigree. To encourage the breeding habits of those who were physically and mentally superior, Galton suggested that a suitable authority issue "eugenic certificates," which would attest to "more than an average share of the several qualities of at least goodness of constitution, of physique, and of mental capacity" (Galton 1905, 23).

12.100. issue . . . in. Include.

12.102. escutcheon. The shield on which a coat of arms is depicted. "A blot on an escutcheon" means a stain on one's reputation or character.

12.109. Dodecagon. The first edition reads "Dodecahedron" instead of "Dodecagon" (see Note 10.27).

12.113. low voice. A quiet voice.

12.114. "an excellent thing in a Woman." An allusion to *King Lear* 5.3. After his youngest daughter Cordelia is hanged, Lear appears on stage holding her dead body in his arms, thinking that he cannot hear her speak because she speaks so quietly:

> Cordelia, Cordelia! stay a little. Ha!
> What is't thou say'st – Her voice was ever soft,
> Gentle and low, an excellent thing in a woman.

115   Such ill-judged marriages are, as might be expected, barren, if they do not result in positive Irregularity or in
diminution of sides; but none of these evils have hitherto
proved sufficiently deterrent. The loss of a few sides in a
highly-developed Polygon is not easily noticed, and is some-
120   times compensated by a successful operation in the Neo-
Therapeutic Gymnasium, as I have described above; and the
Circles are too much disposed to acquiesce in infecundity as
a Law of the superior development. Yet, if this evil be not
arrested, the gradual diminution of the Circular class may
125   soon become more rapid, and the time may be not far distant when, the race being no longer able to produce a Chief
Circle, the Constitution of Flatland must fall.

One other word of warning suggests itself to me, though I
cannot so easily mention a remedy; and this also refers to our
130   relations with Women. About three hundred years ago, it was
decreed by the Chief Circle that, since women are deficient
in Reason but abundant in Emotion, they ought no longer
to be treated as rational, nor receive any mental education.
The consequence was that they were no longer taught to
135   read, nor even to master Arithmetic enough to enable them
to count the angles of their husband or children; and hence
they sensibly declined during each generation in intellectual
power. And this system of female non-education or quietism
still prevails.

140   My fear is that, with the best intentions, this policy has
been carried so far as to react injuriously on the Male Sex.

For the consequence is that, as things now are, we Males
have to lead a kind of bi-lingual, and I may almost say bi-
mental, existence. With Women, we speak of "love," "duty,"
145   "right," "wrong," "pity," "hope," and other irrational and
emotional conceptions, which have no existence, and the
fiction of which has no object except to control feminine
exuberances; but among ourselves, and in our books, we
have an entirely different vocabulary and I may almost say,

12.122. acquiesce in infecundity. Accept unfruitfulness or barrenness. Rosemary Jann neatly observes that the sterility of the Circles' imaginations "is mirrored in their infecundity as a class and spells their defeat in the age to come" (Jann 1985, 487).

12.132. women are deficient in Reason. In *Politics*, Aristotle held that it was natural for the male to rule over the female because the woman's deliberative faculty "is without authority." Genevieve Lloyd traces "women's exclusion from Reason" through the works of Aristotle, Aquinas, Descartes, Rousseau, Kant, Hegel, and Sartre (Lloyd 1984).

12.138. female non-education. For Abbott's efforts on behalf of the education of women, see Appendix B3.

12.138. quietism. Literally, the act of reducing to quietness.

12.143. bi-lingual. Systematic studies of the differences in the use of language by men and women began in the twentieth century, and both the nature and the extent of these differences remain the subject of linguistic research. One theory attributes the differences to male dominance and female subordination in society; another holds that men and women belong to different subcultures and that linguistic differences reflect cultural differences. Either theory might explain the profound differences in male and female speech in Flatland.

150   idiom. "Love" then becomes "the anticipation of benefits";
      "duty" becomes "necessity" or "fitness"; and other words
      are correspondingly transmuted. Moreover, among Women,
      we use language implying the utmost deference for their
      Sex; and they fully believe that the Chief Circle Himself is
155   not more devoutly adored by us than they are: but behind
      their backs they are both regarded and spoken of – by all
      except the very young – as being little better than "mindless
      organisms."

      Our Theology also in the Women's chambers is entirely
160   different from our Theology elsewhere.

      Now my humble fear is that this double training, in lan-
      guage as well as in thought, imposes somewhat too heavy
      a burden upon the young, especially when, at the age of
      three years old, they are taken from the maternal care and
165   taught to unlearn the old language – except for the purpose
      of repeating it in the presence of their Mothers and Nurses –
      and to learn the vocabulary and idiom of Science. Already
      methinks I discern a weakness in the grasp of mathemat-
      ical truth at the present time as compared with the more
170   robust intellect of our ancestors three hundred years ago. I
      say nothing of the possible danger if a Woman should ever
      surreptitiously learn to read and convey to her Sex the result
      of her perusal of a single popular volume; nor of the possibil-
      ity that the indiscretion or disobedience of some infant Male
175   might reveal to a Mother the secrets of the logical dialect.
      On the simple ground of the enfeebling of the Male intel-
      lect, I rest this humble appeal to the highest Authorities to
      reconsider the regulations of Female Education.

12.150. idiom. Language that is peculiar to a limited class of people, where expressions often have meanings different from what is suggested by grammar or logic.

12.168. methinks. It seems to me.

12.172. Woman should ever surreptitiously learn. Another hint that Flatland women are intelligent.

PART II

OTHER WORLDS

" O brave new worlds,
That have such people in them !"

Part II: OTHER WORLDS

Epigraph. The epigraph to Part II is a slight change from Miranda's naive comment on her first sight of the courtly lords and villains in *The Tempest* 5.1. Here "brave" means fine, splendid, or beautiful.

> O, wonder!
> How many goodly creatures are there here!
> How beauteous mankind is! O brave new world,
> That has such people in't.

Aldous Huxley's novel, *Brave New World* (1932), derives its ironic title from this same source.

§13
How I had a Vision of Lineland

IT WAS THE LAST DAY but one of the 1999th year of our era, and the first day of the Long Vacation. Having amused myself till a late hour with my favourite recreation of Geometry, I had retired to rest with an unsolved problem in my
5   mind. In the night I had a dream.

I saw before me a vast multitude of small Straight Lines (which I naturally assumed to be Women) interspersed with other Beings still smaller and of the nature of lustrous Points – all moving to and fro in one and the same Straight
10 Line, and, as nearly as I could judge, with the same velocity.

My view of Lineland

My-self

Women A boy Men My eye Men The KING Men Men A boy Women

The KING's eyes much larger than the reality shewing that HIS MAJESTY could see nothing but a point.

A noise of confused, multitudinous chirping or twittering issued from them at intervals as long as they were moving; but sometimes they ceased from motion, and then all was silence.

Notes on Section 13.

13.2. the Long Vacation. In England, the Long Vacation refers to the summer vacation at the universities and law-courts; at Cambridge, the vacation lasted roughly from mid-June to mid-October. The Long Vacation in Flatland occurs around the beginning of the year, and in this respect, the Flatland calendar is like the Athenian calendar in which the year began, in theory, with the appearance of the first new moon after the summer solstice, and the transition from the old to the new year was celebrated over several days.

13.5. dream. The title of this section refers not to a dream but "a Vision of Lineland." Elsewhere, Abbott calls a "dream" a "night-vision" (Abbott 1907b, 27).

In 1880, the English writer William Hale White (pseudonym Mark Rutherford) wrote a short story in which the narrator, like the Square, has a "geometric" dream. The narrator of White's "A dream of two dimensions," having been frustrated in his attempt to teach "Euclid" to his son, falls unconscious, and passes into a dream universe in which all around him are colored shadows, whereas he is three dimensional. His extra dimension (his intellect) is entirely invisible and incomprehensible to these shadows, particularly his own wife. The story was anonymously printed for private circulation in 1884. White revised it in 1908, and it was first printed for general circulation in White (1915).

15   Approaching one of the largest of what I thought to be
     Women, I accosted her, but received no answer. A second
     and a third appeal on my part were equally ineffectual. Los-
     ing patience at what appeared to me intolerable rudeness, I
     brought my mouth into a position full in front of her mouth
20   so as to intercept her motion, and loudly repeated my ques-
     tion, "Woman, what signifies this concourse, and this strange
     and confused chirping, and this monotonous motion to and
     fro in one and the same Straight Line?"

     "I am no Woman," replied the small Line: "I am the
25   Monarch of the world. But thou, whence intrudest thou into
     my realm of Lineland?" Receiving this abrupt reply, I begged
     pardon if I had in any way startled or molested his Royal
     Highness; and describing myself as a stranger I besought
     the King to give me some account of his dominions. But I
30   had the greatest possible difficulty in obtaining any infor-
     mation on points that really interested me; for the Monarch
     could not refrain from constantly assuming that whatever
     was familiar to him must also be known to me and that I
     was simulating ignorance in jest. However, by persevering
35   questions I elicited the following facts:

     It seemed that this poor ignorant Monarch – as he called
     himself – was persuaded that the Straight Line which he
     called his Kingdom, and in which he passed his existence,
     constituted the whole of the world, and indeed the whole
40   of Space. Not being able either to move or to see, save in
     his Straight Line, he had no conception of anything out of it.
     Though he had heard my voice when I first addressed him,
     the sounds had come to him in a manner so contrary to his
     experience that he had made no answer, "seeing no man," as
45   he expressed it, "and hearing a voice as it were from my own
     intestines." Until the moment when I placed my mouth in his
     World, he had neither seen me, nor heard anything except
     confused sounds beating against – what I called his side, but
     what he called his *inside* or *stomach*; nor had he even now
50   the least conception of the region from which I had come.
     Outside his World, or Line, all was a blank to him; nay, not
     even a blank, for a blank implies Space; say, rather, all was
     non existent.

13.21. concourse. Crowd; literally, a running together.

13.25. thou. In Old English, "thou" and its cases "thee," "thine," and "thy" were used in ordinary speech. By the time of Shakespeare, "you" was being used as for either singular or plural. "Thou" was retained as "the pronoun of (1) affection towards friends, (2) good-humoured superiority to servants, and (3) contempt or anger to strangers. It had, however, already fallen somewhat into disuse, and, being regarded as archaic, was naturally adopted (4) into the higher poetic style and in the language of solemn prayer" (Abbott 1870, 153–154). Here the King of Lineland uses "thou" in sense (3).

13.25. whence intrudest thou. The Square's accounts of his dream of Lineland and the visitation of the Sphere are essentially stories-within-a-story. As in a Shakespearian play-within-a-play, the dialogue in these stories is deliberately stilted.

13.27. molested. Disturbed or put to inconvenience.

13.31. Monarch. "Monarch" is derived from the ancient Greek *mono* (sole or single) and *archen* (ruler); it is an appropriate title for the ruler of a one-dimensional space.

13.51. nay. A word used not simply to deny but to introduce a correction or amplification of what has just been said.

His subjects – of whom the small Lines were Men and the
55   Points Women – were all alike confined in motion and eyesight to that single Straight Line, which was their World. It
need scarcely be added that the whole of their horizon was
limited to a Point; nor could any one ever see anything but a
Point. Man, woman, child, thing – each was a Point to the eye
60   of a Linelander. Only by the sound of the voice could sex or
age be distinguished. Moreover, as each individual occupied
the whole of the narrow path, so to speak, which constituted
his Universe, and no one could move to the right or left
to make way for passers by, it followed that no Linelander
65   could ever pass another. Once neighbours, always neighbours. Neighbourhood with them was like marriage with us.
Neighbours remained neighbours till death did them part.

Such a life, with all vision limited to a Point, and all motion
to a Straight Line, seemed to me inexpressibly dreary; and
70   I was surprised to note the vivacity and cheerfulness of
the King. Wondering whether it was possible, amid circumstances so unfavourable to domestic relations, to enjoy the
pleasures of conjugal union, I hesitated for some time to
question his Royal Highness on so delicate a subject; but at
75   last I plunged into it by abruptly inquiring as to the health
of his family. "My wives and children," he replied, "are well
and happy."

Staggered at this answer – for in the immediate proximity
of the Monarch (as I had noted in my dream before I entered
80   Lineland) there were none but Men – I ventured to reply,
"Pardon me, but I cannot imagine how your Royal Highness
can at any time either see or approach their Majesties, when
there are at least half a dozen intervening individuals, whom
you can neither see through, nor pass by? Is it possible that
85   in Lineland proximity is not necessary for marriage and for
the generation of children?"

"How can you ask so absurd a question?" replied the
Monarch. "If it were indeed as you suggest, the Universe
would soon be depopulated. No, no; neigbourhood is need-
90   less for the union of hearts; and the birth of children is too

13.66. always neighbours. G. T. Fechner was the first person to mention "Lineland" in the literature. In his satirical essay, *"Warum wird die Wurst schief durchschnitten?"* ("Why should the sausage be sliced slantwise?"), he playfully suggests a model of Lineland in which beings might pass through one another.

> "I once sat next to (the mathematician and astronomer August F. Möbius) at a party and asked his advice concerning the composition of a world which had only one dimension instead of three. It seemed to me that such a world had great advantages – it had none of the bothersome complexity of this world and there it would be impossible to go astray. The greatest difficulty seemed to be how people in such a world could change places, and the reader may consider whether he can solve this problem. Together we found two solutions, which seemed entirely practicable. The first was to regard this linear world as bent back on itself forming an ellipse with divine monads as foci. There the people who wished to get past one another would merely reverse their directions and meet half way on the other side. Such a world can be represented naturally as a railroad track, where travel would be rapid, but of course it could accommodate only two people. In the second model, not subject to this limitation, one had to imagine the people as linear waves. As is well known, waves can pass one another without interference, and since our thoughts are already attached to aether waves in the brain, in this setting one being could exchange places with another in reality simultaneously with thinking of doing so" (translated by H. G. Fellner and W. F. Lindgren from Fechner 1875, 398–399).

13.66. like marriage. Prior to the Matrimonial Causes Act of 1857, jurisdiction in English divorce cases lay with the ecclesiastical courts, and a civil divorce required expensive and elaborate legal measures.

important a matter to have been allowed to depend upon
such an accident as proximity. You cannot be ignorant of this.
Yet since you are pleased to affect ignorance, I will instruct
you as if you were the veriest baby in Lineland. Know, then,
95   that marriages are consummated by means of the faculty of
sound and the sense of hearing.

"You are of course aware that every Man has two mouths or
voices – as well as two eyes – a bass at one, and a tenor at the
other, of his extremities. I should not mention this, but that
100   I have been unable to distinguish your tenor in the course
of our conversation." I replied that I had but one voice, and
that I had not been aware that his Royal Highness had two.
"That confirms my impression," said the King, "that you are
not a Man, but a feminine Monstrosity with a bass voice, and
105   an utterly uneducated ear. But to continue.

"Nature having herself ordained that every Man should
wed two wives—" "Why two?" asked I. "You carry your
affected simplicity too far," he cried. "How can there be a
completely harmonious union without the combination of
110   the Four in One, viz. the Bass and Tenor of the Man and the
Soprano and Contralto of the two Women?" "But supposing," said I, "that a man should prefer one wife or three?"
"It is impossible," he said; "it is as inconceivable as that two
and one should make five, or that the human eye should
115   see a Straight Line." I would have interrupted him; but he
proceeded as follows:

"Once in the middle of each week a Law of Nature com-
pels us to move to and fro with a rhythmic motion of more
than usual violence, which continues for the time you would
120   take to count a hundred and one. In the midst of this choral
dance, at the fifty-first pulsation, the inhabitants of the Universe pause in full career, and each individual sends forth his
richest, fullest, sweetest strain. It is in this decisive moment
that all our marriages are made. So exquisite is the adaptation
125   of Bass to Treble, of Tenor to Contralto, that oftentimes the
Loved Ones, though twenty thousand leagues away, recognise at once the responsive note of their destined Lover; and,

13.93. you are pleased to affect ignorance. The phrase "is/are pleased to" means "to have the inclination or disposition for some specified action." In *Flatland*, it is always used sarcastically. Here the King means, "You are pretending to be ignorant."

13.94. veriest. Tiniest.

13.98. two voices. In *Through Nature to Christ*, Abbott describes a case study from Henry Maudsley's *Physiology and Pathology of the Mind*: "Cases have been known where two distinct voices, one bass, representing the moral will, the other falsetto, representing the immoral will, issue in succession from the patient and bear witness to the strife of wills within him" (Abbott 1877a, 448).

13.100. tenor. A pun on two meanings of tenor: the adult male singing voice above baritone and the general sense or meaning of a document or speech.

13.122. in full career. At the height of their activity.

13.125. Treble. Soprano.

13.126. twenty thousand leagues. An allusion to Jules Verne's *20,000 Leagues under the Sea* (1870). The league as a unit of distance has varied with time and place, becoming standardized only recently. In his novel, we can infer that Verne reckoned the nautical league as 2.16 nautical miles. Today, a nautical league is 3 nautical miles, or 5.556 kilometers. According to the *Oxford English Dictionary*, "league" was never in regular use in England but often occurs in poetical or rhetorical statements of distance (Verne, Miller, and Walter 1993).

penetrating the paltry obstacles of distance, Love unites the
three. The marriage in that instant consummated results in
130   a threefold Male and Female offspring which takes its place
in Lineland."

"What! Always threefold?" said I. "Must one wife then
always have twins?"

"Bass-voiced Monstrosity! yes," replied the King. "How
135   else could the balance of the Sexes be maintained, if two
girls were not born for every boy? Would you ignore the
very Alphabet of Nature?" He ceased, speechless for fury;
and some time elapsed before I could induce him to resume
his narrative.

140   "You will not, of course, suppose that every bachelor
among us finds his mates at the first wooing in this universal Marriage Chorus. On the contrary, the process is by
most of us many times repeated. Few are the hearts whose
happy lot it is at once to recognise in each other's voices
145   the partner intended for them by Providence, and to fly into
a reciprocal and perfectly harmonious embrace. With most
of us the courtship is of long duration. The Wooer's voices
may perhaps accord with one of the future wives, but not
with both; or not, at first, with either; or the Soprano and
150   Contralto may not quite harmonise. In such cases Nature
has provided that every weekly Chorus shall bring the three
Lovers into closer harmony. Each trial of voice, each fresh
discovery of discord, almost imperceptibly induces the less
perfect to modify his or her vocal utterance so as to approx-
155   imate to the more perfect. And after many trials and many
approximations, the result is at last achieved. There comes a
day at last, when, while the wonted Marriage Chorus goes
forth from universal Lineland, the three far-off Lovers suddenly find themselves in exact harmony, and, before they
160   are aware, the wedded Triplet is rapt vocally into a duplicate
embrace; and Nature rejoices over one more marriage and
over three more births."

13.137. the very Alphabet of Nature. The fundamental principles of nature. Perhaps an allusion to the *Abecedarium Novum Naturae* (new alphabet of nature), a fragmentary work of Francis Bacon.

13.152. closer harmony. Like Linelanders, mosquitoes use sound for sexual recognition. When a male encounters a female, he opens a "mating ritual" by increasing his wing-beat to emit a higher-frequency tone than the one emitted by the female. In response, the female slightly increases her wing-beat to try to match the tone of the male, while the male slows his wing-beat to match hers. Within seconds, their frequencies are closely matched (Gibson and Russell 2006).

13.155. approximate to. Come close to.

13.157. wonted. Customary, usual. A play on "wanted."

13.158. universal Lineland. The King believes that Lineland includes everything that exists.

13.160. rapt. Carried away. A play on "wrapped."

§14
How I vainly tried to explain the nature of Flatland

THINKING THAT IT WAS TIME to bring down the Monarch
from his raptures to the level of common sense, I determined to endeavour to open up to him some glimpses of
the truth, that is to say of the nature of things in Flatland. So
5   I began thus: "How does your Royal Highness distinguish
the shapes and positions of his subjects? I for my part noticed
by the sense of sight, before I entered your Kingdom, that
some of your people are Lines and others Points, and that
some of the Lines are larger—" "You speak of an impossibil-
10 ity," interrupted the King; "you must have seen a vision; for
to detect the difference between a Line and a Point by the
sense of sight is, as every one knows, in the nature of things,
impossible; but it can be detected by the sense of hearing,
and by the same means my shape can be exactly ascertained.
15 Behold me – I am a Line, the longest in Lineland, over six
inches of Space – "Of Length," I ventured to suggest. "Fool,"
said he, "Space is Length. Interrupt me again, and I have
done."

I apologised; but he continued scornfully, "Since you are
20 impervious to argument, you shall hear with your ears how
by means of my two voices I reveal my shape to my Wives,
who are at this moment six thousand miles seventy yards
two feet eight inches away, the one to the North, the other to
the South. Listen, I call to them."

Notes on Section 14.

Title. Note that the title of Section 14 given in the Contents is "How in my Vision I endeavoured to explain the nature of Flatland, but could not."

14.18. I have done. I am finished (with you).

14.22. six thousand miles. On Earth, sound travels at about 750 miles per hour through the air, and so about 8 hours would be required for sound to travel 6,000 miles. In Lineland, the King's message travels this distance in a moment.

14.23. inches. The King has expressed his own length and the distance between himself and his wives in the English system. The international standard unit of length, a meter, was defined in 1983 as the distance traveled by light in a vacuum in $1/299,792,458$ of a second. An analogous standard unit for Lineland would be the distance traveled by sound in a certain length of time.

25   He chirruped, and then complacently continued: "My
wives at this moment receiving the sound of one of my
voices, closely followed by the other, and perceiving that the
latter reaches them after an interval in which sound can traverse 6.457 inches, infer that one of my mouths is 6.457 inches
30   further from them than the other, and accordingly know my
shape to be 6.457 inches. But you will of course understand
that my Wives do not make this calculation every time they
hear my two voices. They made it, once for all, before we
were married. But they *could* make it at any time. And in
35   the same way I can estimate the shape of any of my Male
subjects by the sense of sound."

"But how," said I, "if a Man feigns a Woman's voice with
one of his two voices, or so disguises his Southern voice that
it cannot be recognised as the echo of the Northern? May not
40   such deceptions cause great inconvenience? And have you
no means of checking frauds of this kind by commanding
your neighbouring subjects to feel one another?" This of
course was a very stupid question, for feeling could not have
answered the purpose; but I asked with the view of irritating
45   the Monarch, and I succeeded perfectly.

"What!" cried he in horror, "explain your meaning." "Feel,
touch, come into contact," I replied. "If you mean by *feeling*,"
said the King, "approaching so close as to leave no space
between two individuals, know, Stranger, that this offence
50   is punishable in my dominions by death. And the reason
is obvious. The frail form of a Woman, being liable to be
shattered by such an approximation, must be preserved by
the State; but since Women cannot be distinguished by the
sense of sight from Men, the Law ordains universally that
55   neither Man nor Woman shall be approached so closely as
to destroy the interval between the approximator and the
approximated.

"And indeed what possible purpose would be served by
this illegal and unnatural excess of approximation which

14.35. I can estimate. The Linelanders' use of sound in ascertaining size and position resembles sonar, a system for use under the sea in which the audible or high-frequency sound reflected or emitted by an object in the sea is used to ascertain its position, nature, or speed.

14.52. approximation. The action of coming near.

60 you call *touching*, when all the ends of so brutal and coarse
a process are attained at once more easily and more exactly
by the sense of hearing? As to your suggested danger of
deception, it is non-existent: for the Voice, being the essence
of one's Being, cannot be thus changed at will. But come,
65 suppose that I had the power of passing through solid things,
so that I could penetrate my subjects, one after another, even
to the number of a billion, verifying the size and distance
of each by the sense of *feeling*: how much time and energy
would be wasted in this clumsy and inaccurate method!
70 Whereas now, in one moment of audition, I take as it were the
census and statistics, local, corporeal, mental, and spiritual,
of every living being in Lineland. Hark, only hark!"

So saying he paused and listened, as if in an ecstasy, to a
sound which seemed to me no better than a tiny chirping
75 from an innumerable multitude of lilliputian grasshoppers.

"Truly," replied I, "your sense of hearing serves you in
good stead, and fills up many of your deficiencies. But permit
me to point out that your life in Lineland must be deplorably
dull. To see nothing but a Point! Not even to be able to
80 contemplate a Straight Line! Nay, not even to know what
a Straight Line is! To see, yet to be cut off from those Linear prospects which are vouchsafed to us in Flatland! Better
surely to have no sense of sight at all than to see so little! I
grant you I have not your discriminative faculty of hearing;
85 for the concert of all Lineland which gives you such intense
pleasure, is to me no better than a multitudinous twittering
or chirping. But at least I can discern, by sight, a Line from a
Point. And let me prove it. Just before I came into your kingdom, I saw you dancing from left to right, and then from
90 right to left, with seven Men and a Woman in your immediate proximity on the left, and eight Men and two Women on
your right. Is not this correct?"

"It is correct," said the King, "so far as the numbers and
sexes are concerned, though I know not what you mean by

14.67. billion. Originally in Great Britain a million millions.

14.70. audition. The action of listening or hearing.

14.72 Hark, only hark! Listen, just listen!

14.73. ecstasy. A state of being beside one's self.

14.75. lilliputian. Lilliput is the name of an imaginary country in Jonathan Swift's *Gulliver's Travels* (1726). Lilliput is peopled by pygmies "not six inches high"; hence, "lilliputian" means small or petty. Like *Flatland*, *Gulliver's Travels* was a pseudonymous satire.

14.82. vouchsafe. To grant or bestow, especially by a superior to an inferior.

95 'right' and 'left.' But I deny that you saw these things. For how could you see the Line, that is to say the inside, of any Man? But you must have heard these things, and then dreamed that you saw them. And let me ask what you mean by those words 'left' and 'right.' I suppose it is your way of
100 saying Northward and Southward."

"Not so," replied I; "besides your motion of Northward and Southward, there is another motion which I call from right to left."

King. Exhibit to me, if you please, this motion from left to
105 right.

I. Nay, that I cannot do, unless you could step out of your Line altogether.

King. Out of my Line? Do you mean out of the world? Out of Space?

110 *I.* Well, yes. Out of *your* World. Out of *your* Space. For your Space is not the true Space. True Space is a Plane; but your Space is only a Line.

King. If you cannot indicate this motion from left to right by yourself moving in it, then I beg you to describe it to me
115 in words.

I. If you cannot tell your right side from your left, I fear that no words of mine can make my meaning clear to you. But surely you cannot be ignorant of so simple a distinction.

King. I do not in the least understand you.

120 *I.* Alas! How shall I make it clear? When you move straight on, does it not sometimes occur to you that you *could* move in some other way, turning your eye round so as to look in the direction towards which your side is now fronting? In other words, instead of always moving in the direction of
125 one of your extremities, do you never feel a desire to move in the direction, so to speak, of your side?

14.104. if you please. A sarcastic phrase suggesting that the King does not believe that the Square can demonstrate left-to-right movement.

14.111. your Space is not the true Space. Albert Einstein points out that the words "red," "hard," and "disappointed" are unlikely to be misinterpreted because they are connected with our elementary experiences. "But in the case of words such as 'place' or 'space' whose relation with psychological experience is less direct, there exists a far-reaching uncertainty of interpretation" (Foreword to Jammer 1969, xii).

King. Never. And what do you mean? How can a man's inside "front" in any direction? Or how can a man move in the direction of his inside?

130   *I.* Well then, since words cannot explain the matter, I will try deeds, and will move gradually out of Lineland in the direction which I desire to indicate to you.

At the word I began to move my body out of Lineland. As long as any part of me remained in his dominion and in his
135   view, the King kept exclaiming, "I see you, I see you still;

you are not moving." But when I had at last moved myself out of his Line, he cried in his shrillest voice, "She is vanished; she is dead." "I am not dead," replied I; "I am simply out of
140   Lineland, that is to say, out of the Straight Line which you call Space, and in the true Space, where I can see things as they are. And at this moment I can see your Line, or side – or inside as you are pleased to call it; and I can see also the Men and Women on the North and South of you, whom I
145   will now enumerate, describing their order, their size, and the interval between each."

When I had done this at great length, I cried triumphantly, "Does this at last convince you?" And, with that, I once more entered Lineland, taking up the same position as before.

150   But the Monarch replied, "If you were a Man of sense – though, as you appear to have only one voice I have little doubt you are not a Man but a Woman – but, if you had a particle of sense, you would listen to reason. You ask me to believe that there is another Line besides that which my

14.148 "Does this at last convince you?" The Square might have tried lifting the King into two-dimensional space and setting him back down with his bass and tenor voices reversed. Such a relocation might not have convinced the King of the existence of space outside Lineland, but it would certainly have left him unsettled and without any way of regaining his original orientation. If we pick up a Flatland woman and set her back down in the reverse direction, she will no doubt be unnerved by the experience, but she can regain her original orientation by rotating 180°. (See Note 16.103 for analogous "reversals" in three- and four-dimensional space.)

155 senses indicate, and another motion besides that of which
I am daily conscious. I, in return, ask you to describe in
words or indicate by motion that other Line of which you
speak. Instead of moving, you merely exercise some magic
art of vanishing and returning to sight; and instead of any
160 lucid description of your new World, you simply tell me the
numbers and sizes of some forty of my retinue, facts known
to any child in my capital. Can anything be more irrational
or audacious? Acknowledge your folly or depart from my
dominions."

165   Furious at his perversity, and especially indignant that he
professed to be ignorant of my Sex, I retorted in no measured
terms, "Besotted Being! You think yourself the perfection of
existence, while you are in reality the most imperfect and
imbecile. You profess to see, whereas you can see nothing
170 but a Point! You plume yourself on inferring the existence of
a Straight Line; but I *can see* Straight Lines, and infer the existence of Angles, Triangles, Squares, Pentagons, Hexagons,
and even Circles. Why waste more words? Suffice it that I
am the completion of your incomplete self. You are a Line,
175 but I am a Line of Lines, called in my country a Square:
and even I, infinitely superior though I am to you, am of
little account among the great Nobles of Flatland, whence
I have come to visit you, in the hope of enlightening your
ignorance."

180   Hearing these words the King advanced towards me with
a menacing cry as if to pierce me through the diagonal; and
in that same moment there arose from myriads of his subjects
a multitudinous war-cry, increasing in vehemence till at last
methought it rivalled the roar of an army of a hundred thou-
185 sand Isosceles, and the artillery of a thousand Pentagons.
Spell-bound and motionless, I could neither speak nor move
to avert the impending destruction; and still the noise grew
louder, and the King came closer, when I awoke to find the
breakfast-bell recalling me to the realities of Flatland.

14.167. in no measured terms. "Measured terms" means language that is carefully considered, deliberate, and restrained.

14.167. Besotted. Intellectually stupefied.

14.170. plume yourself. To congratulate oneself or show self-satisfaction, especially regarding something trivial or ridiculous.

14.175. Line of Lines. The Square means that he may be regarded as the union or aggregate of a collection of line segments, as Figure 14.1 illustrates. Every point of the line segment AB corresponds to a line segment (namely, the one that contains the point and is parallel to AD), and every point of the square ABCD is on one of these lines. In the same way, we might regard a cube as a line of squares and a sphere as a line of circular discs (see Note 15.107).

Figure 14.1. A square as a line of lines.

The Italian mathematician Bonaventura Cavalieri expresses this notion in his best-known work, *The Geometry of Indivisibles* (1635), where he asserts that a line is made up of an infinite number of points, a surface of an infinite number of lines, and a solid of an infinite number of planes.

14.184. methought. It seemed to me.

14.185. artillery of a thousand Pentagons. This is a mistake – it is the Squares who are the artillerymen.

14.189. breakfast-bell. The custom of holding morning prayers was a common practice throughout the Victorian upper and middle classes. The entire household was gathered together "at the same hour every day punctually summoned by the ringing of a bell or gong before the adult breakfast was served" (Davidoff 1973, 35).

§15
Concerning a Stranger from Spaceland

FROM DREAMS I PROCEED TO facts.

It was the last day of the 1999th year of our era. The pattering of the rain had long ago announced nightfall; and I was sitting[1] in the company of my wife, musing on the events of the past and the prospects of the coming year, the coming century, the coming Millennium.

My four Sons and two orphan Grandchildren had retired to their several apartments; and my Wife alone remained with me to see the old Millennium out and the new one in.

I was rapt in thought, pondering in my mind some words that had casually issued from the mouth of my youngest Grandson, a most promising young Hexagon of unusual brilliancy and perfect angularity. His uncles and I had been giving him his usual practical lesson in Sight Recognition, turning ourselves upon our centres, now rapidly, now more slowly, and questioning him as to our positions; and his answers had been so satisfactory that I had been induced to reward him by giving him a few hints on Arithmetic as applied to Geometry.

---

[1] When I say "sitting," of course I do not mean any change of attitude such as you in Spaceland signify by that word; for as we have no feet, we can no more "sit" nor "stand" (in your sense of the word) than one of your soles or flounders.

Nevertheless, we perfectly well recognise the different mental states of volition implied in "lying," "sitting," and "standing," which are to some extent indicated to a beholder by a slight increase of lustre corresponding to the increase of volition.

But on this, and a thousand other kindred subjects, time forbids me to dwell.

Notes on Section 15.

15.3. rain . . . announced nightfall. The occurrence of rain in Flatland is so regular that it serves as a "sunset" to mark the close of each day.

15.4. my wife. For Abbott's wife, see Appendix B1, 1863.

15.6. Millennium. A period of one thousand years. According to one interpretation of Revelation 20:1–5, the period of one thousand years during which Christ will reign in person on Earth.

15.7. My four Sons. For Abbott's only son, Edwin, see Appendix B1, 1868.

15.16. questioning him as to our positions. In "static" sight recognition, one observes the varying brightness of a stationary figure's edges to determine the size of its angles. Here the boy already knows that his grandfather is a square and his uncles are pentagons; he is using "kinetic" sight recognition to find their positions as they rotate. In Kinetic Sight Recognition, a Flatlander identifies a figure by walking around it and observing the size of the visual angle from all directions (equivalently, by observing the changing visual angle as the figure rotates). More formally, for a figure K inside a circle C and having the same center of gravity as C, we define the angle function of K along C to be the function that assigns to each point x of C the measure of the visual angle of K at x. János Kincses proved that every convex polygon is determined by its angle function. That is, if P_1 and P_2 are convex polygons that have the same angle function along some circle, then $P_1 = P_2$ (Kincses 2003).

Figure 15.1 is the graph of an angle function of a pentagon. It depicts the changing visual angle of a stationary observer watching a pentagon rotate 180°. As the number of sides of a regular polygon increases, the angle function becomes flatter, and every angle function of a circle is constant. For the angle function of a woman, see Figure 15.2.

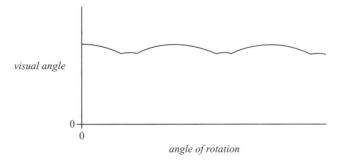

visual angle

angle of rotation

Figure 15.1 An angle function of a pentagon.

Footnote. you in Spaceland. This corrects the first edition, which reads "you in Flatland."

Footnote. soles or flounders. To illustrate how a curvature of space might give an illusion of attractive force, the physicist Arthur Eddington relates a fable of a race of flat fish swimming in curved paths around a mound in the ocean floor – a mound they cannot see because they are two-dimensional (Eddington 1921, 95–96).

20   Taking nine Squares, each an inch every way, I had put them together so as to make one large Square, with a side of three inches, and I had hence proved to my little Grandson that – though it was impossible for us to *see* the inside of the Square – yet we might ascertain the number of square inches
25   in a Square by simply squaring the number of inches in the side: "and thus," said I, "we know that 3^2, or 9, represents the number of square inches in a Square whose side is 3 inches long."

The little Hexagon meditated on this a while and then said
30   to me: "But you have been teaching me to raise numbers to the third power; I suppose 3^3 must mean something in Geometry; what does it mean?" "Nothing at all," replied I, "not at least in Geometry; for Geometry has only Two Dimensions." And then I began to show the boy how a Point
35   by moving through a length of three inches makes a Line of three inches, which may be represented by 3; and how a Line of three inches, moving parallel to itself through a length of three inches, makes a Square of three inches every way, which may be represented by 3^2.

40   Upon this, my Grandson, again returning to his former suggestion, took me up rather suddenly and exclaimed, "Well, then, if a Point by moving three inches, makes a Line of three inches represented by 3; and if a straight Line of three inches, moving parallel to itself, makes a Square of
45   three inches every way, represented by 3^2; it must be that a Square of three inches every way, moving somehow parallel to itself (but I don't see how) must make a Something else (but I don't see what) of three inches every way – and this must be represented by 3^3."

50   "Go to bed," said I, a little ruffled by his interruption: "if you would talk less nonsense, you would remember more sense."

So my Grandson had disappeared in disgrace; and there I sat by my Wife's side, endeavouring to form a retrospect
55   of the year 1999 and of the possibilities of the year 2000, but

15.24. see the inside of the Square. From our vantage point above Flatland, we can see the entire form of its inhabitants and buildings. Lacking this perspective, the Square has learned the shapes of Flatland objects by moving around their perimeters and observing their sides and angles.

15.31. third power ... 3^3. Unlike a Flatlander, we would read "3^3" as "three cubed." In the nineteenth century, the fourth power of a number or quantity was called the biquadrate, and the fifth power the sursolid.

15.34. Geometry has only Two Dimensions. A one-point space is said to be zero-dimensional because no motion is possible. A space like Lineland in which only one motion is possible (north/south) is said to be one-dimensional. Flatland is two-dimensional because any motion in Flatland can be achieved by combining two motions (north/south and east/west) and no fewer than two motions will suffice. Spaceland is three-dimensional because every motion in Spaceland can be achieved by combining three motions (north/south, east/west, and up/down) and no fewer than three motions will suffice. Two-dimensionality is nicely illustrated by an Etch A Sketch™, which was a popular toy in the 1960s. The underside of the Etch A Sketch™ glass screen is coated with a mixture of aluminum powder and plastic beads. The left and right knobs control the horizontal and vertical rods, which move the stylus located at the intersection of the two rods. As the stylus moves, it scrapes the powder from the screen leaving behind a dark "curve." By twisting both knobs simultaneously, one can draw any possible curve on the two-dimensional surface (Rucker 1984, chapter 1).

"We can see the characteristic features of dimensions most clearly when we scale an object up or down. Consider the problem of preparing a photograph for mailing. A square photograph requires a certain amount of string and a certain amount of wrapping paper. If we double the size of the square photograph, we need twice as much string and four times as much paper. Doubling the size of a cubical box requires twice as much string, four times as much paper, and eight times as much packing material. Similarly, if we double the size of an entrance hall, then all linear quantities, like the length of wiring, are doubled. But the quantities involving area, like the amount of paint for the walls and the square feet of carpeting for the floor, are multiplied by four, and quantities involving volume, like the cubic feet of space to be handled by the air conditioning units, increase by a factor of eight" (Banchoff 1990a, 13–14).

15.38. moving parallel to itself ... makes a Square. The Square ought to say "moving perpendicular to itself." In general, a line segment moving parallel to itself in a constant direction generates a parallelogram.

15.41. took me up. Interrupted me.

not quite able to shake off the thoughts suggested by the prattle of my bright little Hexagon. Only a few sands now remained in the half-hour glass. Rousing myself from my reverie I turned the glass Northward for the last time in the old Millennium; and in the act, I exclaimed aloud, "The boy is a fool."

60

Straightway I became conscious of a Presence in the room, and a chilling breath thrilled through my very being. "He is no such thing," cried my Wife, "and you are breaking the Commandments in thus dishonouring your own Grandson." But I took no notice of her. Looking round in every direction I could see nothing; yet still I *felt* a Presence, and shivered as the cold whisper came again. I started up. "What is the matter?" said my Wife, "there is no draught; what are you looking for? There is nothing." There was nothing; and I resumed my seat, again exclaiming, "The boy is a fool, I say; 3^3 can have no meaning in Geometry." At once there came a distinctly audible reply, "The boy is not a fool; and 3^3 has an obvious Geometrical meaning."

65

70

My Wife as well as myself heard the words, although she did not understand their meaning, and both of us sprang forward in the direction of the sound. What was our horror when we saw before us a Figure! At the first glance it appeared to be a Woman, seen sideways; but a moment's observation shewed me that the extremities passed into dimness too rapidly to represent one of the Female Sex; and I should have thought it a Circle, only that it seemed to change its size in a manner impossible for a Circle or for any regular Figure of which I had had experience.

75

80

But my Wife had not my experience, nor the coolness necessary to note these characteristics. With the usual hastiness and unreasoning jealousy of her Sex, she flew at once to the conclusion that a Woman had entered the house through some small aperture. "How comes this person here?" she exclaimed, "you promised me, my dear, that there should be no ventilators in our new house." "Nor are there any," said I; "but what makes you think that the stranger is a Woman?

85

90

15.58. half-hour glass. Through the eighteenth century, time was kept at sea by means of a half-hour sandglass, which consisted of two sealed glass vessels connected by a narrow neck, and containing just enough sand as took one half-hour to drain from the uppermost vessel into the lower. One possible shape of the Square's two-dimensional half-hour glass is a figure shaped like the numeral "8" with an appropriate hole to permit the drainage of two-dimensional "sand."

15.63. thrilled through my very being. Compare *Romeo and Juliet* 4.3: "I have a faint cold fear thrills through my veins."

15.68. started up. Arose suddenly.

15.74. 3³ has an obvious Geometrical meaning. The "obvious" geometrical meaning of 3^3 is by no means obvious to any Flatlander. This passage is the first of several that reveal the Sphere's shortcomings as a teacher.

15.75. myself. Abbott makes an ungrammatical use of "myself" to create euphony: "My wife as well as myself heard the words."

15.77. What was our horror. How great was our horror. Abbott uses this construction to create a double consonance: what was . . . when we.

15.79. it appeared to be a Woman, seen sideways. As the Sphere moves up or down, the size of his circular cross-section (his intersection with Flatland) changes continuously. The Square and his wife "see" this changing cross-section as a line segment varying in length. The only explanation they have for such variation in apparent length is that they are witnessing a rotating figure, and the only figure for which the apparent length varies markedly as it rotates is a woman. The graphs below illustrate how difficult it would be to distinguish between a rotating woman and a sphere passing through Flatland.

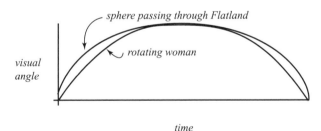

Figure 15.2. The angle functions of a sphere passing through Flatland and a rotating woman.

I see by my power of Sight Recognition – " "Oh, I have no
patience with your Sight Recognition," replied she, " 'Feeling
95   is believing' and 'A Straight Line to the touch is worth a Circle
to the sight' " – two Proverbs, very common with the Frailer
Sex in Flatland.

"Well," said I, for I was afraid of irritating her, "if it must be
so, demand an introduction." Assuming her most gracious
100   manner, my Wife advanced towards the Stranger, "Permit
me, Madam, to feel and be felt by—" then, suddenly recoil-
ing, "Oh! it is not a Woman, and there are no angles either,
not a trace of one. Can it be that I have so misbehaved to a
perfect Circle?"

105   "I am indeed, in a certain sense a Circle," replied the Voice,
"and a more perfect Circle than any in Flatland; but to speak
more accurately, I am many Circles in one." Then he added
more mildly, "I have a message, dear Madam, to your husband, which I must not deliver in your presence; and, if
110   you would suffer us to retire for a few minutes—" But my
Wife would not listen to the proposal that our august Visitor
should so incommode himself, and assuring the Circle that
the hour of her own retirement had long passed, with many
reiterated apologies for her recent indiscretion, she at last
115   retreated to her apartment.

I glanced at the half-hour glass. The last sands had fallen.
The second Millennium had begun.

15.95. "Feeling" is believing. An allusion to John 20:25: "But (Thomas) said unto them, Except I shall see in his hands the print of the nails, and put my finger into the print of the nails, and thrust my hand into his side, I will not believe." Abbott discusses this passage in Abbott (1917, 710 ff).

15.103. not a trace. A play on two meanings of trace: a minute amount and the intersection of a line or surface with a surface.

15.107. many Circles in one. Figure 15.3 illustrates what the Sphere means when he says that he is "many circles in One." Just as a square is a "line of lines," a sphere is a "line of circular discs." That is, every point of the line segment joining the top and bottom of the sphere corresponds to exactly one circular disc, and every point of the sphere belongs to one of these discs.

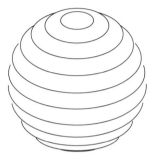

Figure 15.3. The Sphere as "many Circles in one."

15.109. I must not deliver in your presence. The Stranger has chosen the Square to be "initiated into the mysteries," which were to be kept secret from the uninitiated. He repeats his insistence that no one other than the Square may witness his appearance at line 17.76.

15.110. suffer us to retire. Allow us to withdraw to a place of seclusion.

15.112. incommode. Inconvenience.

15.117. second Millennium. This ought to be "third Millennium." The error was corrected in the 1926 Blackwell edition.

§16
How the Stranger vainly endeavoured to reveal to me in words the mysteries of Spaceland

As SOON AS THE SOUND of the Peace-cry of my departing Wife had died away, I began to approach the Stranger with the intention of taking a nearer view and of bidding him be seated: but his appearance struck me dumb and motionless
5   with astonishment. Without the slightest symptoms of angularity he nevertheless varied every instant with gradations of size and brightness scarcely possible for any Figure within the scope of my experience. The thought flashed across me that I might have before me a burglar or cut-throat, some
10 monstrous Irregular Isosceles, who, by feigning the voice of a Circle, had obtained admission somehow into the house, and was now preparing to stab me with his acute angle.

In a sitting-room, the absence of Fog (and the season happened to be remarkably dry), made it difficult for me to trust
15 to Sight Recognition, especially at the short distance at which I was standing. Desperate with fear, I rushed forward with an unceremonious, "You must permit me, Sir—" and felt him. My Wife was right. There was not the trace of an angle, not the slightest roughness or inequality: never in my life
20 had I met with a more perfect Circle. He remained motionless while I walked round him, beginning from his eye and returning to it again. Circular he was throughout, a perfectly satisfactory Circle; there could not be a doubt of it. Then followed a dialogue, which I will endeavour to set down as
25 near as I can recollect it, omitting only some of my profuse apologies – for I was covered with shame and humiliation that I, a Square, should have been guilty of the impertinence

Notes on Section 16.

16.2. the Peace-cry of my departing Wife. The first edition reads, "the sound of my Wife's retreating footsteps." For the origin of this change, see Appendix A2, footnote 10.

16.5. without the slightest symptoms of angularity. Having no perceptible corners.

of feeling a Circle. It was commenced by the Stranger
with some impatience at the lengthiness of my introductory
30   process.

Stranger. Have you felt me enough by this time? Are you
not introduced to me yet?

I. Most illustrious Sir, excuse my awkwardness, which
arises not from ignorance of the usages of polite society, but
35   from a little surprise and nervousness, consequent on this
somewhat unexpected visit. And I beseech you to reveal my
indiscretion to no one, and especially not to my Wife. But
before your Lordship enters into further communications,
would he deign to satisfy the curiosity of one who would
40   gladly know whence his Visitor came?

Stranger. From Space, from Space, Sir: whence else?

I. Pardon me, my Lord, but is not your Lordship already
in Space, your Lordship and his humble servant, even at this
moment?

45   *Stranger.* Pooh! what do you know of Space? Define Space.

I. Space, my Lord, is height and breadth indefinitely
prolonged.

Stranger. Exactly: you see you do not even know what
Space is. You think it is of Two Dimensions only; but I
50   have come to announce to you a Third – height, breadth,
and length.

I. Your Lordship is pleased to be merry. We also speak of
length and height, or breadth and thickness, thus denoting
Two Dimensions by four names.

55   *Stranger.* But I mean not only three names, but Three
Dimensions.

I. Would your Lordship indicate or explain to me in what
direction is the Third Dimension, unknown to me?

Stranger. I came from it. It is up above and down below.

16.43. humble servant. A phrase used formally in addressing a person regarded as one's superior; also a mode of expressing submission to another's opinion.

16.45. Define Space. Albert Einstein observes that it is profoundly difficult to say what space is and how it is related to matter and motion: "[W]e entirely shun the vague word 'space,' of which, we must honestly acknowledge, we cannot form the slightest conception, and we replace it by 'motion relative to a practically rigid body of reference'" (Einstein 1921, 9–10).

16.52. is pleased to be merry. Is inclined to be facetious.

60   *I.* My Lord means seemingly that it is Northward and Southward.

Stranger. I mean nothing of the kind. I mean a direction in which you cannot look, because you have no eye in your side.

65   *I.* Pardon me, my Lord, a moment's inspection will convince your Lordship that I have a perfect luminary at the juncture of two of my sides.

Stranger. Yes: but in order to see into Space you ought to have an eye, not on your Perimeter, but on your side, that
70   is, on what you would probably call your inside; but we in Spaceland should call it your side.

I. An eye in my inside! An eye in my stomach! Your Lordship jests.

Stranger. I am in no jesting humour. I tell you that I come
75   from Space, or, since you will not understand what Space means, from the Land of Three Dimensions whence I but lately looked down upon your Plane which you call Space forsooth. From that position of advantage I discerned all that you speak of as *solid* (by which you mean "enclosed on four
80   sides"), your houses, your churches, your very chests and safes, yes even your insides and stomachs, all lying open and exposed to my view.

I. Such assertions are easily made, my Lord.

Stranger. But not easily proved, you mean. But I mean to
85   prove mine.

When I descended here, I saw your four Sons, the Pentagons, each in his apartment, and your two Grandsons the Hexagons; I saw your youngest Hexagon remain a while with you and then retire to his room, leaving you and your
90   Wife alone. I saw your Isosceles servants, three in number,

16.66. I have a perfect luminary. The Square refers to his eye as a luminary, that is, a light-giving organ. The extramission theory of vision reached its full development with Plato, who describes it in the *Timaeus*, 45b–d. According to this theory, a stream of light or fire emanates from the observer's eye and coalesces with daylight to form a "single homogeneous body," which serves as the material intermediary between the visible object and the eye (Lindberg 1976, 3–6).

16.74. humour. Mood or temperament without reference to anything amusing. This usage derives from the ancient notion that the body contained four fluids or "humours" whose relative proportions determined a person's health and temperament.

16.78. forsooth. In truth, truly. "Forsooth" had become obsolete before *Flatland* appeared and was often used to imply contempt, as it is here.

in the kitchen at supper, and the little Page in the scullery. Then I came here, and how do you think I came?

I. Through the roof, I suppose.

Stranger. Not so. Your roof, as you know very well, has
95   been recently repaired, and has no aperture by which even a Woman could penetrate. I tell you I come from Space. Are you not convinced by what I have told you of your children and household?

I. Your Lordship must be aware that such facts touching the
100   belongings of his humble servant might be easily ascertained by any one in the neighbourhood possessing your Lordship's ample means of obtaining information.

Stranger. (*To himself*). What must I do? Stay; one more argument suggests itself to me. When you see a Straight
105   Line – your wife, for example – how many Dimensions do you attribute to her?

I. Your Lordship would treat me as if I were one of the vulgar who, being ignorant of Mathematics, suppose that a Woman is really a Straight Line, and only of One Dimen-
110   sion. No, no, my Lord; we Squares are better advised, and are as well aware as your Lordship that a Woman, though popularly called a Straight Line, is, really and scientifically, a very thin Parallelogram, possessing Two Dimensions, like the rest of us, viz., length and breadth (or thickness).

115   *Stranger.* But the very fact that a Line is visible implies that it possesses yet another Dimension.

I. My Lord, I have just acknowledged that a Woman is broad as well as long. We see her length, we infer her breadth; which, though very slight, is capable of measurement.

120   *Stranger.* You do not understand me. I mean that when you see a Woman, you ought – besides inferring her breadth – to see her length, and to *see* what we call her *height*; although that last Dimension is infinitesimal in your country. If a Line were mere length without "height," it would cease to occupy

16.91. scullery. A small room attached to a kitchen in which the washing of dishes and other dirty work is done; a back kitchen.

16.99. touching. As regards; concerning.

16.103. What must I do? As the Square uses the words "left" and "right," it seems that he can distinguish between the two vertices adjacent to the one where his eye/mouth is located. Let L denote the vertex on the west side and R the one on the east side as he faces north. His "back" vertex B is the one diagonally opposite his eye E. If the Sphere were to lift the Square out of Flatland and return him "upside-down," he would appear as the "mirror-image" of his former self – when he faced north, vertex R would be on the west side and L on the east side. Although such a mirror-reversal would not convince the Square of the existence of a third dimension, he would soon learn that no motion in the plane of Flatland would restore him to his original orientation.

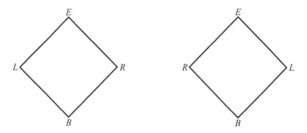

Figure 16.1. The Square and his mirror image.

H. G. Wells's "The Plattner Story" (1896) describes an analogous "reversal" in four-dimensional space. A science teacher, Gottfried Plattner, explodes a "greenish" powder that blows him into four-dimensional space. After nine days in this universe, Plattner slips on a boulder, falls, and the bottle containing the remaining powder is smashed and explodes. In another moment, he finds himself back on Earth, where he discovers that his heart is now on his right side; further, it seems that his entire body has had its left and right sides transposed (Wells 1952).

For the geometric notion of orientability, see the essay "Immanuel Kant and nonorientability" (Banchoff 1990a, 192–193), Rucker (1984, chapter 4), and Burger (1965). See also William Sleator's *The Boy Who Reversed Himself*.

16.103. Lines 16.103 through 16.143 as well as "(To himself.) I can do neither." in line 16.144 were added to the first edition. For the origin of these changes, see Appendix A2.

16.113. a Straight Line, is, really and scientifically, a very thin Parallelogram. The seventeenth-century Oxford theologian and mathematician John Wallis declared that a parallelogram whose altitude is infinitely small or zero as "scarcely anything but a line" (Boyer 1949, 170–171).

125   space and would become invisible. Surely you must recognize this?

I. I must indeed confess that I do not in the least understand your Lordship. When we in Flatland see a Line, we see
length and *brightness*. If the brightness disappears, the line
130   is extinguished, and, as you say, ceases to occupy space. But
am I to suppose that your Lordship gives to brightness the
title of a Dimension, and that what we call "bright" you call
"high"?

Stranger. No, indeed. By "height" I mean a Dimension like
135   your length: only, with you, "height" is not so easily perceptible, being extremely small.

I. My Lord, your assertion is easily put to the test. You say
I have a Third Dimension, which you call "height." Now,
Dimension implies direction and measurement. Do but mea-
140   sure my "height," or merely indicate to me the direction in
which my "height" extends, and I will become your convert.
Otherwise, your Lordship's own understanding must hold
me excused.

Stranger. (*To himself*). I can do neither. How shall I convince
145   him? Surely a plain statement of facts followed by ocular
demonstration ought to suffice. – Now, Sir; listen to me.

You are living on a Plane. What you style Flatland is the
vast level surface of what I may call a fluid, on, or in, the top
of which you and your countrymen move about, without
150   rising above it or falling below it.

I am not a plane Figure, but a Solid. You call me a Circle;
but in reality I am not a Circle, but an infinite number of
Circles, of size varying from a Point to a Circle of thirteen
inches in diameter, one placed on the top of the other. When
155   I cut through your plane as I am now doing, I make in your
plane a section which you, very rightly, call a Circle. For even
a Sphere – which is my proper name in my own country – if
he manifest himself at all to an inhabitant of Flatland – must
needs manifest himself as a Circle.

16.125. If a Line were mere length without "height"... **invisible.** Abbott put this fallacious assertion into the mouth of the Sphere to illustrate his fallibility. For the origin of this change to the first edition, see Appendix A2, Footnote 2.

16.132. gives to brightness the title of Dimension. The Square is asking whether the Sphere intends the word "dimension" in a figurative sense to mean an aspect or attribute of an entity.

16.147. What you style Flatland. What you call Flatland. This use of the word "Flatland" is at odds with the Square's earlier statement that he was calling his world Flatland not because that is what he and his fellow countrymen call it but to make its nature clearer to us, his readers.

16.148. level surface of... **a fluid.** The marine biologist A. E. Walsby discovered tiny, flat, transparent square bacteria floating at the surface of brine pools in the Sinai Peninsula (Walsby 1980).

16.156. section. Cross-section. The writer Paul Lake uses a dimensional metaphor to describe "the shape of poetry." He maintains that the shape of a poem is not its two-dimensional outline on a page, which he likens to the cross-section of a higher-dimensional object. He says that the essence of a poem is the "four-dimensional shape it creates when spoken or read" (Lake 2001, 166).

16.157. Sphere. A. K. Dewdney observes that the geometry of the participants "serves to neutralize the human content" of the encounter between the Sphere and the Square, and so the reader is not forced to interpret the event as a spiritual or religious experience. "But the vibration is there and is reinforced enough times in the course of *Flatland*'s telling to leave no doubt that the metaphysical dimension is Abbott's main interest" (Dewdney 1984a, 10).

160   Do you not remember – for I, who see all things, discerned
last night the phantasmal vision of Lineland written upon
your brain – do you not remember, I say, how, when you
entered the realm of Lineland, you were compelled to manifest yourself to the King, not as a Square, but as a Line,
165 because that Linear Realm had not Dimensions enough to
represent the whole of you, but only a slice or section of you?
In precisely the same way, your country of Two Dimensions
is not spacious enough to represent me, a being of Three, but
can only exhibit a slice or section of me, which is what you
170 call a Circle.

The diminished brightness of your eye indicates
incredulity. But now prepare to receive proof positive of the
truth of my assertions. You cannot indeed see more than one
of my sections, or Circles, at a time; for you have no power
175 to raise your eye out of the plane of Flatland; but you can
at least see that, as I rise in Space, so my sections become
smaller. See now, I will rise; and the effect upon your eye
will be that my Circle will become smaller and smaller till it
dwindles to a point and finally vanishes.

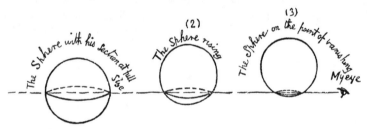

180   There was no "rising" that I could see; but he diminished
and finally vanished. I winked once or twice to make sure
that I was not dreaming. But it was no dream. For from the
depths of nowhere came forth a hollow voice – close to my
heart it seemed – "Am I quite gone? Are you convinced now?
185 Well, now I will gradually return to Flatland and you shall
see my section become larger and larger."

Every reader in Spaceland will easily understand that my
mysterious Guest was speaking the language of truth and
even of simplicity. But to me, proficient though I was in
190 Flatland Mathematics, it was by no means a simple matter.
The rough diagram given above will make it clear to any

16.160. Do you not remember. C. Howard Hinton suggests that a two-dimensional being might be led to conceive of the existence of a third dimension of space if he should happen to imagine a being confined to a line.

> If he (a two-dimensional being) were to imagine a being confined to a single straight line, he might realise that he himself could move in two directions, while the creature in a straight line could only move in one. Having made this reflection he might ask, 'But why is the number of directions limited to two? Why should there not be three?' (Hinton 1880, 18)

Abbott has given the Square an extended dream encounter with a one-dimensional being, but he has not provided the Square with enough imagination to interpret this dream, which foreshadows the "appearance" in Flatland of a being from three-dimensional space. Further, the Square is so constrained by Flatland thinking that he is oblivious to the analogy between his own attempt to convince the King of Lineland of the existence of a second dimension of space and the Sphere's attempt to convince him of the existence of a third dimension of space.

16.178. my Circle will become smaller. The Square experiences the passage of the Sphere through Flatland as a circular disc expanding and contracting with time.

Illustration. In a perspective view, the sphere's circular cross-sections appear to be elliptical, as in Figure 16.2; however, in the Square's badly drawn diagram, they resemble the cross-section of a double-convex lens.

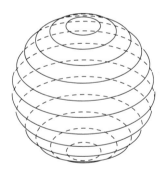

Figure 16.2. Cross-sections of a sphere.

16.183. hollow voice. A reminiscence of *Hamlet* 1.5, where Shakespeare has the ghost speak to Hamlet and Horatio from beneath the stage.

Spaceland child that the Sphere, ascending in the three positions indicated there, must needs have manifested himself to me, or to any Flatlander, as a Circle, at first of full size, then
195 small, and at last very small indeed, approaching to a Point. But to me, although I saw the facts before me, the causes were as dark as ever. All that I could comprehend was, that the Circle had made himself smaller and vanished, and that he had now reappeared and was rapidly making himself larger.

200 When he regained his original size, he heaved a deep sigh; for he perceived by my silence that I had altogether failed to comprehend him. And indeed I was now inclining to the belief that he must be no Circle at all, but some extremely clever juggler; or else that the old wives' tales were true,
205 and that after all there were such people as Enchanters and Magicians.

After a long pause he muttered to himself, "One resource alone remains, if I am not to resort to action. I must try the method of Analogy." Then followed a still longer silence,
210 after which he continued our dialogue.

Sphere. Tell me, Mr. Mathematician; if a Point moves Northward, and leaves a luminous wake, what name would you give to the wake?

I. A straight Line.

215 *Sphere.* And a straight Line has how many extremities?

I. Two.

Sphere. Now conceive the Northward straight Line moving parallel to itself, East and West, so that every point in it leaves behind it the wake of a straight Line. What name will you
220 give to the Figure thereby formed? We will suppose that it moves through a distance equal to the original straight Line. – What name, I say?

I. A Square.

16.197. All that I could comprehend. In addition to the challenge of imagining higher-dimensional spaces, *Flatland* introduces the reader to the task of trying to imagine a two-dimensional creature imagining a three-dimensional object. We cannot know the conception of space of two-dimensional beings. Because we have no knowledge of their sense perceptions, even the assumption that their idea of space is derived from a sense of "touch or vision" is insufficient to warrant any conclusion about their spatial conceptions. However, the governing principle of *Flatland* is that Flatlanders are "like us," and so for these particular two-dimensional figures, we may reason from our own experience to try to imagine what conception of space they may have. This kind of analogical reasoning is a common approach to understanding our own relation to four-dimensional space. [continued]

16.204. juggler. One who deceives by trickery.

16.204. the old wives' tales. The ascription to "old wives" of "tales" that perpetuate ancient superstitions began in antiquity. Plato mentions them in the *Gorgias*, and in his first letter to Timothy, Paul warns, "Refuse profane and old wives' fables."

16.209. Analogy. In its broadest sense, an analogy is a comparison between two things in which several of the attributes of one thing correspond to attributes of the other. Analogy comes from the Greek word *analogia*, defined by Aristotle as an equality of ratios or proportion: A is to B as C is to D (*Nicomachean Ethics* 5.3). The definitive essay on Plato's extensive use of analogy is chapter 12 of Richard Robinson's *Plato's Earlier Dialectic*. [continued]

16.211. Mr. Mathematician. The Stranger is referring sarcastically to the Square's assertion that, unlike the common people in Flatland, he is not ignorant of mathematics.

16.211. if a Point moves Northward. Figure 16.4 illustrates the derivation sequence: A point (a 0-cube) moving in a constant direction determines a line segment (a 1-cube); a segment moving perpendicular to itself in a plane determines a square (a 2-cube); a square moving perpendicular to itself in three-dimensional space determines a cube (a 3-cube). At line 19.183, the Square conjectures the existence of the next term of this sequence, the 4-cube or hypercube, which he calls an Extra-Cube.

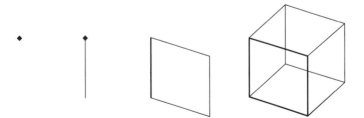

Figure 16.4. The derivation sequence.

[continued]

16.218. East and West. East to West.

Sphere. And how many sides has a Square? How many
225 angles?

I. Four sides and four angles.

Sphere. Now stretch your imagination a little, and conceive
a Square in Flatland, moving parallel to itself upward.

I. What? Northward?

230   *Sphere.* No, not Northward; upward; out of Flatland
altogether.

If it moved Northward, the Southern points in the Square
would have to move through the positions previously occupied by the Northern points. But that is not my meaning.

235   I mean that every Point in you – for you are a Square and
will serve the purpose of my illustration – every Point in you,
that is to say in what you call your inside, is to pass upwards
through Space in such a way that no Point shall pass through
the position previously occupied by any other Point; but each
240 Point shall describe a straight Line of its own. This is all in
accordance with Analogy; surely it must be clear to you.

Restraining my impatience – for I was now under a strong
temptation to rush blindly at my Visitor and to precipitate
him into Space, or out of Flatland, anywhere, so that I could
245 get rid of him – I replied:—

"And what may be the nature of the Figure which I am to
shape out by this motion which you are pleased to denote
by the word 'upward'? I presume it is describable in the
language of Flatland."

16.243. precipitate. To throw violently. It is not at all clear what the Square means by "into Space, or out of Flatland" because for him Flatland is all of "space."

250   *Sphere.* Oh, certainly. It is all plain and simple, and in strict
accordance with Analogy – only, by the way, you must not
speak of the result as being a Figure, but as a Solid. But I will
describe it to you. Or rather not I, but Analogy.

We began with a single Point, which of course – being itself
255   a Point – has only *one* terminal Point.

One Point produces a Line with *two* terminal Points.

One Line produces a Square with *four* terminal Points.

Now you can give yourself the answer to your own question: 1, 2, 4, are evidently in Geometrical Progression. What
260   is the next number?

I. Eight.

Sphere. Exactly. The one Square produces a *Something-
which-you-do-not-as-yet-know-a-name-for-but-which-we-call-a-
Cube* with *eight* terminal Points. Now are you convinced?

265   *I.* And has this Creature sides, as well as angles or what
you call "terminal Points?"

Sphere. Of course; and all according to Analogy. But, by
the way, not what *you* call sides, but what *we* call sides. You
would call them *solids*.

270   *I.* And how many solids or sides will appertain to this
Being whom I am to generate by the motion of my inside in
an "upward" direction, and whom you call a Cube?

Sphere. How can you ask? And you a mathematician! The
side of anything is always, if I may so say, one Dimension
275   behind the thing. Consequently, as there is no Dimension
behind a Point, a Point has 0 sides; a Line, if I may say, has
2 sides (for the Points of a line may be called by courtesy, its
sides); a Square has 4 sides; 0, 2, 4; what Progression do you
call that?

280   *I.* Arithmetical.

16.259. Geometrical Progression. The sequence (1, 2, 4, 8 . . .) is an example of a geometric progression – a sequence in which each term after the first is equal to the product of the preceding one and a fixed number. When a line segment moves perpendicular to itself to form a square, the initial line segment and the final line segment both have two vertices, so the square has four vertices. When a square moves perpendicular to itself to form a cube, the initial square and the final square both have four vertices, so the cube has eight vertices. In general, each time an n-cube moves perpendicular to itself to form an $(n + 1)$-cube, the number of vertices doubles because the initial n-cube and final n-cube have the same number of vertices.

16.270. appertain. Belong to.

16.275. The side . . . one Dimension behind. The sides of a line segment are endpoints, the sides of a square are edges, the sides of a cube are squares, the sides of a hypercube are cubes, and in general the sides of an n-cube are $(n - 1)$-cubes.

Mathematicians have extended in various ways the idea that a line is one-dimensional, a plane is two-dimensional, and space is three-dimensional to a much more general notion of dimension. The Square's observation that "the side of anything is always . . . one Dimension behind the thing" contains the essence of one such extended definition, the "small inductive" dimension given independently by Karl Menger and P. J. Urysohn in 1922 (Menger 1943).

16.280. Arithmetical. The sequence of nonnegative even integers (0, 2, 4, 6 . . .) is an example of an arithmetic progression – a sequence where the terms are equally spaced. That is, each term after the first is equal to the sum of the preceding one and a fixed number. As a line segment moves perpendicular to itself to form a square, each of its two sides (vertices) moves to form a line segment. These two line segments together with the initial and final segments are the four sides of the square. As a square moves perpendicular to itself to form a cube, each of its four sides (edges) moves to form a square. These four squares together with the initial and final squares are the six sides of the cube. In general, each time an n-cube moves perpendicular to itself to form an $(n + 1)$-cube, the number of "sides" increases by two.

Sphere. And what is the next number?

I. Six.

Sphere. Exactly. Then you see you have answered your own
question. The Cube which you will generate will be bounded
285 by six sides, that is to say, six of your insides. You see it all
now, eh?

"Monster," I shrieked, "be thou juggler, enchanter, dream,
or devil, no more will I endure thy mockeries. Either thou or
I must perish." And saying these words I precipitated myself
290 upon him.

§17
How the Sphere, having in vain
tried words, resorted to deeds

IT WAS IN VAIN. I brought my hardest right angle into
violent collision with the Stranger, pressing on him with a
force sufficient to have destroyed any ordinary Circle: but
I could feel him slowly and unarrestably slipping from my
5   contact; not edging to the right nor to the left, but moving
somehow out of the world, and vanishing to nothing. Soon
there was a blank. But still I heard the Intruder's voice.

Sphere. Why will you refuse to listen to reason? I had hoped
to find in you – as being a man of sense and an accomplished
10   mathematician – a fit apostle for the Gospel of the Three
Dimensions, which I am allowed to preach once only in a
thousand years: but now I know not how to convince you.
Stay, I have it. Deeds, and not words, shall proclaim the truth.
Listen, my friend.

16.288. mockeries. The Square is not convinced of the existence of a cube, nor should he be. The Sphere's "method of Analogy" suggests that the number of terminal points of the objects in the derivation sequence form a geometric sequence (1, 2, 4, 8, 16...), and the number of sides of these objects form an arithmetic sequence (0, 2, 4, 6, 8...). As we observed previously, these conjectures about certain properties of higher-dimensional mathematical objects are readily proved; however, they are of no use whatever in demonstrating the physical existence of such objects.

Of course, Abbott was well aware of the limitations of arguments from analogy: "The Argument from Analogy therefore, so far as it is an argument at all, comes under the head of Induction. Otherwise it is not an argument, but a metaphorical illustration of an argument" (Abbott and Seeley 1871, 273).

Notes on Section 17.

17.10. apostle. The chief advocate of a new principle or system; a person sent with mandates by another. From the Greek *apostolos*, one sent forth.

17.10. the Gospel. From the Old English *godspel*, the good tidings.

17.12. once only in a thousand years. It is not clear what authority restricts the Sphere to preaching just once each millennium.

17.13. Deeds, and not words. This ancient proverb is the moral of Aesop's, "The Boasting Traveler."

15   I have told you I can see from my position in Space the
inside of all things that you consider closed. For example, I
see in yonder cupboard near which you are standing, several
of what you call boxes (but like everything else in Flatland,
they have no tops nor bottoms) full of money; I see also two
20   tablets of accounts. I am about to descend into that cupboard
and to bring you one of those tablets. I saw you lock the
cupboard half an hour ago, and I know you have the key in
your possession. But I descend from Space; the doors, you
see, remain unmoved. Now I am in the cupboard and am
25   taking the tablet. Now I have it. Now I ascend with it.

I rushed to the closet and dashed the door open. One of
the tablets was gone. With a mocking laugh, the Stranger
appeared in the other corner of the room, and at the same
time the tablet appeared upon the floor. I took it up. There
30   could be no doubt – it was the missing tablet.

I groaned with horror, doubting whether I was not out of
my senses; but the Stranger continued: "Surely you must
now see that my explanation, and no other, suits the phenomena. What you call Solid things are really superficial;
35   what you call Space is really nothing but a great Plane. I am
in Space, and look down upon the insides of the things of
which you only see the outsides. You could leave this Plane
yourself, if you could but summon up the necessary volition.
A slight upward or downward motion would enable you to
40   see all that I can see."

"The higher I mount, and the further I go from your Plane,
the more I can see, though of course I see it on a smaller scale.
For example, I am ascending; now I can see your neighbour
the Hexagon and his family in their several apartments; now
45   I see the inside of the Theatre, ten doors off, from which the
audience is only just departing; and on the other side a Circle
in his study, sitting at his books. Now I shall come back to
you. And, as a crowning proof, what do you say to my giving
you a touch, just the least touch, in your stomach? It will not
50   seriously injure you, and the slight pain you may suffer cannot be compared with the mental benefit you will receive."

17.20. tablets of accounts. For ordinary purposes such as letters and accounts, the Greeks used thin wooden tablets coated with wax. Such tablets provided reusable, portable writing surfaces in antiquity and throughout the Middle Ages.

17.25. taking the tablet. The Sphere can easily remove an object from the Square's cupboard without opening its doors because the cupboard is open in the direction perpendicular to Flatland. Similarly, a four-dimensional being could remove an object from a closed, three-dimensional safe because it has no wall facing the direction of the fourth dimension (Rucker 1984, 25–28).

17.34. suits the phenomena. A hypothesis is said to suit (or save) the phenomena provided that it satisfactorily explains the observed facts. "Though the phrase does not occur in those of Plato's works that remain, *sozein ta phainomena* appears to have its origin in the Platonic tradition, where saving the appearances is what astronomers are said to bring about by means of the mathematical combinations of the circular motions that Plato supposedly laid down as canonical" (McMullin 2003, 458).

17.34. superficial. Two-dimensional.

17.48. crowning proof. A play on two meanings of "to crown": to add the finishing touch and to bring to a successful conclusion.

Before I could utter a word of remonstrance, I felt a shooting pain in my inside, and a demoniacal laugh seemed to issue from within me. A moment afterwards the sharp agony
55 had ceased, leaving nothing but a dull ache behind, and the Stranger began to reappear, saying, as he gradually increased in size, "There, I have not hurt you much, have I? If you are not convinced now, I don't know what will convince you. What say you?"

60 My resolution was taken. It seemed intolerable that I should endure existence subject to the arbitrary visitations of a Magician who could thus play tricks with one's very stomach. If only I could in any way manage to pin him against the wall till help came!

65 Once more I dashed my hardest angle against him, at the same time alarming the whole household by my cries for aid. I believe, at the moment of my onset, the Stranger had sunk below our Plane, and really found difficulty in rising. In any case he remained motionless, while I, hearing, as I thought,
70 the sound of some help approaching, pressed against him with redoubled vigour, and continued to shout for assistance.

A convulsive shudder ran through the Sphere. "This must not be," I thought I heard him say: "either he must listen to reason, or I must have recourse to the last resource of civi-
75 lization." Then, addressing me in a louder tone, he hurriedly exclaimed, "Listen: no stranger must witness what you have witnessed. Send your Wife back at once, before she enters the apartment. The Gospel of Three Dimensions must not be thus frustrated. Not thus must the fruits of one thousand
80 years of waiting be thrown away. I hear her coming. Back! back! Away from me, or you must go with me – whither you know not – into the Land of Three Dimensions!"

"Fool! Madman! Irregular!" I exclaimed; "never will I release thee; thou shalt pay the penalty of thine impostures."

85 "Ha! Is it come to this?" thundered the Stranger: "then meet your fate: out of your Plane you go. Once, twice, thrice! 'Tis done!"

17.74. I must have recourse to the last resource of civilization. I must turn to the only remaining means of enlightenment.

17.81. Away. Get away.

17.85. Is it come to this? In Germanic languages generally, including Old English, the perfect of certain verbs was formed not with the verb "to have" but with the verb "to be." These were intransitive verbs, typically expressing the condition or state attained rather than the action of reaching it (Denison 1993, 344). Such usage is common in the King James Bible: "I am come that they might have life"; "He is not here: for he is risen, as he said." The Square uses it again at line 18.72: "Behold, I am become as a God."

17.86. out of your Plane. As his life in Flatland has rendered the Square incapable of imagining even the possibility of another spatial dimension, the Sphere's appeals to his reason and previous experience have proved to be fruitless. The philosophical journey requires complete "detachment" from the world, and so the Sphere lifts the Square out of his plane into space.

17.87 'Tis done! A common phrase in the works of Shakespeare, used most memorably in *Macbeth* 1.7: "If it were done when 'tis done, then 'twere well/It were done quickly."

§18
How I came to Spaceland,
and what I saw there

AN UNSPEAKABLE HORROR SEIZED ME. There was a darkness; then a dizzy, sickening sensation of sight that was not
like seeing; I saw a Line that was no Line; Space that was
not Space: I was myself, and not myself. When I could find
5   voice, I shrieked aloud in agony, "Either this is madness or it
is Hell." "It is neither," calmly replied the voice of the Sphere,
"it is Knowledge; it is Three Dimensions: open your eye once
again and try to look steadily."

I looked, and, behold, a new world! There stood before me,
10   visibly incorporate, all that I had before inferred, conjectured,
dreamed, of perfect Circular beauty. What seemed the centre
of the Stranger's form lay open to my view: yet I could see no
heart, nor lungs, nor arteries, only a beautiful harmonious
Something – for which I had no words; but you, my Readers
15   in Spaceland, would call it the surface of the Sphere.

Prostrating myself mentally before my Guide, I cried,
"How is it, O divine ideal of consummate loveliness and
wisdom that I see thy inside, and yet cannot discern thy
heart, thy lungs, thy arteries, thy liver?" "What you think
20   you see, you see not," he replied; "it is not given to you, nor
to any other Being to behold my internal parts. I am of a different order of Beings, from those in Flatland. Were I a Circle,
you could discern my intestines, but I am a Being, composed
as I told you before, of many Circles, the Many in the One,
25   called in this country a Sphere. And, just as the outside of
a Cube is a Square, so the outside of a Sphere presents the
appearance of a Circle."

Notes on Section 18.

18.3. I saw a Line that was no Line. In a letter to J. B. Priestley, H. G. Wells discusses the Square's ability to see while above Flatland. "I was started (on fourth-dimension) by A. Square & 'Flatland.' It's too intricate a question to argue in a letter, but I think you can get at the catch in the business, if you think what again must have happened to a Square lifted out of his two dimensional universe. It must remain <u>flat</u>. He would not see his former plane as A. Square assumed. He would simply be in another plane. If the latter were inclined to the former, he would see the former only as a linear trace (without detail) on the latter" (Wells 1937).

As Wells says, the Square sees only a series of one-dimensional images of Flatland; nevertheless, he may be able to process these images to form a complete two-dimensional mental image. There are animals that have essentially one-dimensional retinas, and such animals acquire an image of three-dimensional objects by a "scanning" process. "[J]umping spiders and carnivorous sea snails have a linear curved retina consisting only of a band of photoreceptors no wider than 6–7 photoreceptors. This is, in essence, a one-dimensional retina consisting of a line of photoreceptors. Nevertheless, the jumping spiders, at least, have surprisingly good vision and must have some form of three-dimensional processing as they are most successful at prey capture. These creatures swing their retinas through an arc using the principle of scanning for acquisition of the image" (Schwab 2004, 988).

18.6 "Either this is madness or it is Hell." Like the *Republic*, 515d–516a, where Plato describes the temporary blindness and confusion of the prisoner who has been dragged out of the cave into the sunlight, this passage is a figure for the disorientation and anguish experienced by a person whose false conception of knowledge has been exposed.

18.10. incorporate. United in one body. Perhaps the Square is thinking of the circular cross-sections of the Sphere, which he observed when the Sphere visited Flatland, as being united to form a Sphere.

18.16. Guide. Plato indicates the difficulty of the journey out of the cave by having another person drag the prisoner forcibly up the ascent that leads out of the cave into the light of the sun (*Republic*, 515e). The Sphere, who is the agent of the Square's "journey," resembles a *daimon*, a supernatural being in classical Greek literature often credited with assigning their destiny to mortals. Plato describes a *daimon* as a being intermediate between the gods and men who serves as a kind of spirit-guide.

18.24. the Many in the One. Early Greek philosophy was dominated by the problem of "the one and the many." The problem was formulated in various ways, for example: In what sense are the world and our knowledge of it one? For Plato, it was the problem of understanding the relationship between a single Form and its many particulars or instances. For the Square witnessing a three-dimensional solid pass through Flatland, it is the problem of trying to form a conception of the (one) solid based on the (many) images of two-dimensional cross-sections that he observes.

Bewildered though I was by my Teacher's enigmatic utterance, I no longer chafed against it, but worshipped him in
30   silent adoration. He continued, with more mildness in his
voice. "Distress not yourself if you cannot at first understand the deeper mysteries of Spaceland. By degrees they
will dawn upon you. Let us begin by casting back a glance at
the region whence you came. Return with me a while to the
35   plains of Flatland, and I will show you that which you have
often reasoned and thought about, but never seen with the
sense of sight – a visible angle." "Impossible!" I cried; but,
the Sphere leading the way, I followed as if in a dream, till
once more his voice arrested me: "Look yonder, and behold
40   your own Pentagonal house, and all its inmates."

I looked below, and saw with my physical eye all that
domestic individuality which I had hitherto merely inferred
with the understanding. And how poor and shadowy
was the inferred conjecture in comparison with the reality
45   which I now beheld! My four Sons calmly asleep in the

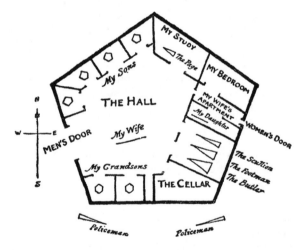

North-Western rooms, my two orphan Grandsons to the
South; the Servants, the Butler, my Daughter, all in their
several apartments. Only my affectionate Wife, alarmed by
my continued absence, had quitted her room and was roving
50   up and down in the Hall, anxiously awaiting my return. Also

18.40. inmates. Inhabitants. In early use, the word meant lodgers or subtenants.

18.45. the reality which I now beheld. What is significant about the Square's viewpoint in three-dimensional space is not the ability to see the insides of objects in Flatland but the comprehensive view that the position affords him. Plato says that having achieved *synoptikos* ("seeing the whole together" or "taking a comprehensive view") is the defining attribute of the dialectical nature (*Republic*, 537c).

Illustration. If we assume that the illustration is drawn to scale and that the Square's wife is 1 foot long, then each side of his house is about 4 feet long and its area is about 27.5 square feet. A Flatland male occupies approximately the same floor space as a human, and so this house must be very crowded.

A scullion is a domestic servant of the lowest rank in a household who performed the menial offices of the kitchen. Flatlanders have no feet, and so it is curious that they have a footman, a male servant whose duties are to attend the door, the carriage, and the table.

18.47. my Daughter. For Abbott's only daughter, Mary, see Appendix B1, 1870.

18.48. affectionate. From the context the obsolete meaning, "mentally affected" (by her husband's continued absence), seems to be what is intended.

the Page, aroused by my cries, had left his room, and under pretext of ascertaining whether I had fallen somewhere in a faint, was prying into the cabinet in my study. All this I could now *see*, not merely infer; and as we came nearer and
55   nearer, I could discern even the contents of my cabinet, and the two chests of gold and the tablets of which the Sphere had made mention.

Touched by my Wife's distress, I would have sprung downward to reassure her, but I found myself incapable of motion.
60   "Trouble not yourself about your Wife," said my Guide: "she will not be long left in anxiety; meantime let us take a survey of Flatland."

Once more I felt myself rising through space. It was even as the Sphere had said. The further we receded from the
65   object we beheld, the larger became the field of vision. My native city, with the interior of every house and every creature therein, lay open to my view in miniature. We mounted higher, and lo, the secrets of the earth, the depths of mines and inmost caverns of the hills, were bared before me.

70   Awestruck at the sight of the mysteries of the earth, thus unveiled before my unworthy eye, I said to my Companion, "Behold, I am become as a God. For the wise men in our country say that to see all things, or as they express it, *omnividence*, is the attribute of God alone." There was something
75   of scorn in the voice of my Teacher as he made answer: "Is it so indeed? Then the very pick-pockets and cut-throats of my country are to be worshipped by your wise men as being Gods: for there is not one of them that does not see as much as you see now. But trust me, your wise men are wrong."

80   *I.* Then is omnividence the attribute of others besides Gods?

Sphere. I do not know. But, if a pick-pocket or a cut-throat of our country can see everything that is in your country, surely that is no reason why the pick-pocket or cut-throat
85   should be accepted by you as a God. This omnividence, as

18.69. secrets of the earth ... mines and inmost caverns. The great limestone caves, which are common in Greece, are often associated with the mysteries.

18.72. I am become as a God. Perhaps an allusion to Genesis 3:5, where the serpent says to Eve, "your eyes shall be opened, and ye shall be as gods."

18.76. pick-pockets and cut-throats. Abbott returns to this illustration in *The Spirit on the Waters*, where he posits a four-dimensional "Super-solid" that would be to us what we are to Flatlanders. "He, then, would be to us what some among us might be disposed to call the All-seeing and Omnipresent God. But no Christian ought to be able – it is perhaps too much to say '*is* able' – to give the name of God to a Super-solid, who may perhaps be a wholly despicable creature, an escaped convict from the four-dimensional land" (Abbott 1897, 29–33).

you call it – it is not a common word in Spaceland – does it
make you more just, more merciful, less selfish, more loving?
Not in the least. Then how does it make you more divine?

I. "More merciful, more loving!" But these are the qualities
90   of women! And we know that a Circle is a higher Being than
a Straight Line, in so far as knowledge and wisdom are more
to be esteemed than mere affection.

Sphere. It is not for me to classify human faculties according
to merit. Yet many of the best and wisest in Spaceland think
95   more of the affections than of the understanding, more of
your despised Straight Lines than of your belauded Circles.
But enough of this. Look yonder. Do you know that building?

I looked, and afar off I saw an immense Polygonal structure, in which I recognized the General Assembly Hall of
100 the States of Flatland, surrounded by dense lines of Pentagonal buildings at right angles to each other, which I knew to
be streets; and I perceived that I was approaching the great
Metropolis.

"Here we descend," said my Guide. It was now morning,
105 the first hour of the first day of the two thousandth year of
our era. Acting, as was their wont, in strict accordance with
precedent, the highest Circles of the realm were meeting in
solemn conclave, as they had met on the first hour of the first
day of the year 1000, and also on the first hour of the first
110 day of the year 0.

The minutes of the previous meetings were now read by
one whom I at once recognized as my brother, a perfectly
Symmetrical Square, and the Chief Clerk of the High Council. It was found recorded on each occasion that: "Whereas
115 the States had been troubled by divers ill-intentioned persons pretending to have received revelations from another
World, and professing to produce demonstrations whereby
they had instigated to frenzy both themselves and others,
it had been for this cause unanimously resolved by the
120 Grand Council that on the first day of each millenary, special

18.90. the qualities of women! According to the Sphere, justice, mercy, unselfish-ness, and love are divine attributes; yet the Square scorns these qualities as womanly. Abbott is satirizing the contemporary view that such moral qualities are exclusively feminine.

18.103. Metropolis. In England, "the metropolis" refers to London as a whole, as distinct from "the city," that part within the ancient boundaries.

18.106. wont. Custom, habit.

18.108. conclave. A formal, close assembly, especially of a religious character.

18.110. year 0. Note that in our calendar 1 AD follows 1 BC.

18.112. my brother. For Abbott's brother Sydney, see Appendix B1, 1887.

18.113. perfectly Symmetrical Square. The Greek word *symmetria* meant the har-mony or proportion of the constituent parts of an object. The notion of symmetry is made precise in mathematics by defining various types of symmetry. Two familiar types are bilateral symmetry, which is conspicuous in the human body, and rotational symmetry, which is most completely exhibited by a circle. A figure has bilateral symmetry if there is a line along which a mirror can be placed to reflect either half of the figure so that it reproduces the other half. A figure has n-fold rotational symmetry if rotating it through an angle of $(360 / n)°$ leaves it coincident with its original position. A square has four lines of symmetry (the two lines through its opposite corners and the two lines through the midpoints of its opposite edges) and four-fold rotational symmetry. In general, every regular polygon with n sides has n lines of symmetry and n-fold rotational symmetry.

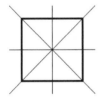

Figure 18.1. The lines of symmetry of a square.

18.113. High Council. The High Council in Flatland corresponds roughly to the council of the *Areopagus,* an aristocratic body in ancient Athens with a member-ship derived from those who had held high public office.

18.115. divers. Two meanings are possible. The first, although somewhat archaic in ordinary usage, is well known in legal and scriptural phraseology: an indef-inite numeral word expressing multiplicity without committing the speaker to "many" or "few." A second meaning, now obsolete, fits the context very well: differing from or opposed to what is right, good, or profitable; perverse, evil.

18.116. revelations. Knowledge revealed to man by a divine or supernatural agency. Flatland's Priests are empiricists; they deny the possibility of acquiring knowledge except through experience and any mention of a revelation they treat as heresy.

18.120. millenary. A thousandth anniversary or its celebration; a millennium.

injunctions be sent to the Prefects in the several districts
of Flatland, to make strict search for such misguided persons, and without formality of mathematical examination, to
destroy all such as were Isosceles of any degree, to scourge
125 and imprison any regular Triangle, to cause any Square or
Pentagon to be sent to the district Asylum, and to arrest any
one of higher rank, sending him straightway to the Capital
to be examined and judged by the Council."

"You hear your fate," said the Sphere to me, while the
130 Council was passing for the third time the formal resolution.
"Death or imprisonment awaits the Apostle of the Gospel
of Three Dimensions." "Not so," replied I, "the matter is
now so clear to me, the nature of real space so palpable, that
methinks I could make a child understand it. Permit me but
135 to descend at this moment and enlighten them." "Not yet,"
said my Guide, "the time will come for that. Meantime I must
perform my mission. Stay thou there in thy place." Saying
these words, he leaped with great dexterity into the sea (if
I may so call it) of Flatland, right in the midst of the ring of
140 Counsellors. "I come," cried he, "to proclaim that there is a
land of Three Dimensions."

I could see many of the younger Counsellors start back
in manifest horror, as the Sphere's circular section widened
before them. But on a sign from the presiding Circle – who
145 showed not the slightest alarm or surprise – six Isosceles of a
low type from six different quarters rushed upon the Sphere.
"We have him," they cried; "No; yes; we have him still! he's
going! he's gone!"

"My Lords," said the President to the Junior Circles of
150 the Council, "there is not the slightest need for surprise; the
secret archives, to which I alone have access, tell me that
a similar occurrence happened on the last two millennial
commencements. You will, of course, say nothing of these
trifles outside the Cabinet."

155 Raising his voice, he now summoned the guard. "Arrest
the policemen; gag them. You know your duty." After he had

18.121. Prefect. Governor.

18.124. scourge. To punish severely by whipping. Scourging was an ancient form of purification intended to drive out evil.

18.129. You hear your fate. Whether the Sphere knows exactly what the Square's fate is, it is a fate for which he (the Sphere) is directly responsible. This "assignment of fate" supports our conjecture that Abbott's model for the Sphere is a Greek *daimon*.

18.133. palpable. Readily perceived by the mind.

18.137. my mission. The purpose of this "mission" is not obvious. Earlier, the Sphere insisted that no stranger may witness what the Square was about to witness, yet here he swooped down upon the meeting of the Council in what was certainly not a serious attempt to convince its members of the existence of three-dimensional space. Predictably, all he achieved was the extermination of a few expendable Isosceles policemen and the imprisonment of the Square's brother. A few of the younger Circles were momentarily taken aback by the appearance of the Sphere's circular cross-section, but the President's assurances soon quieted their alarm.

The imprisonment of the Square's brother was the only consequential outcome of this "mission," and it may be that bringing about this imprisonment was just what the Sphere intended. Perhaps the Sphere anticipated the Square's subsequent imprisonment and wanted his brother in prison with him to provide the Square with affirmation that his "initiation into the mysteries" actually happened. Support for this conjecture is found at line 19.6, where the Sphere alludes to the Square and his brother being together in prison.

18.140. Counsellor. A member of the Council. Since the sixteenth century, the word has been spelled "councillor."

18.143. manifest horror. A play on two meanings of "manifest": obvious horror and horror at the manifestation.

18.146. quarters. Directions.

18.148. We have him . . . he's gone! A reminiscence of *Hamlet* 1.1, where the soldiers rush upon the ghost of Hamlet's father: "Bernardo: 'Tis here! / Horatio: 'Tis here! / Marcellus: 'Tis gone!'"

consigned to their fate the wretched policemen – ill-fated and
unwilling witnesses of a State-secret which they were not to
be permitted to reveal – he again addressed the Counsellors.
160   "My Lords, the business of the Council being concluded, I
have only to wish you a happy New Year." Before departing,
he expressed, at some length, to the Clerk, my excellent but
most unfortunate brother, his sincere regret that, in accordance with precedent and for the sake of secrecy, he must
165   condemn him to perpetual imprisonment, but added his satisfaction that, unless some mention were made by him of
that day's incident, his life would be spared.

§19
How, though the Sphere showed me other mysteries of Spaceland, I still desired more; and what came of it

WHEN I SAW MY POOR brother led away to imprisonment, I
attempted to leap down into the Council Chamber, desiring
to intercede on his behalf, or at least bid him farewell. But I
found that I had no motion of my own. I absolutely depended
5   on the volition of my Guide, who said in gloomy tones,
"Heed not thy brother; haply thou shalt have ample time
hereafter to condole with him. Follow me."

18.166. added his satisfaction. Said that he was satisfied.

Notes on Section 19.

19.6. haply. By chance.

19.7. ample time hereafter to condole with him. The Sphere anticipates that the Square will join his brother in prison. Note that "condole" means to grieve with; not to be confused with "console," which means "to comfort."

Once more we ascended into space. "Hitherto," said the Sphere, "I have shown you naught save Plane Figures and

10

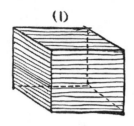

their interiors. Now I must introduce you to Solids, and reveal to you the plan upon which they are constructed. Behold this multitude of moveable square cards. See, I put one on another,

15

not, as you supposed, Northward of the other, but *on* the other. Now a second, now a third. See, I am building up a Solid by a multitude of Squares parallel to one another. Now the Solid is com-

20

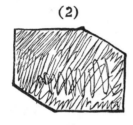

plete, being as high as it is long and broad, and we call it a Cube."

"Pardon me, my Lord," replied I; "but to my eye the appearance is as of an Irregular Figure whose inside is

25

laid open to the view; in other words, methinks I see no Solid, but a Plane such as we infer in Flatland; only of an Irregularity which betokens some monstrous criminal, so that the very sight of it is painful to my eyes."

30   "True," said the Sphere; "it appears to you a Plane, because you are not accustomed to light and shade and perspective; just as in Flatland a Hexagon would appear a Straight Line to one who has not the Art of Sight Recognition. But in reality it is a Solid, as you shall learn by the sense of Feeling."

35   He then introduced me to the Cube, and I found that this marvellous Being was indeed no plane, but a Solid; and that he was endowed with six plane sides and eight terminal points called solid angles; and I remembered the saying of the Sphere that just such a Creature as this would be formed by

19.24. Irregular Figure. We readers recognize Figure 19.1 not as an irregular hexagon but as the projection of a cube onto the two-dimensional surface of the page. The projections of six of the edges of the cube form a hexagon containing distorted images of six squares, the bounding faces of the cube.

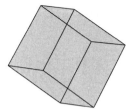

Figure 19.1. An "irregular figure," the projection of a cube.

19.29. painful to my eyes. A small slip; the Square has only one eye.

40   a Square moving, in Space, parallel to himself: and I rejoiced
to think that so insignificant a Creature as I could in some
sense be called the Progenitor of so illustrious an offspring.

But still I could not fully understand the meaning of what
my Teacher had told me concerning "light" and "shade" and
45   "perspective"; and I did not hesitate to put my difficulties
before him.

Were I to give the Sphere's explanation of these matters,
succinct and clear though it was, it would be tedious to an
inhabitant of Space, who knows these things already. Suffice
50   it, that by his lucid statements, and by changing the position
of objects and lights, and by allowing me to feel the several
objects and even his own sacred Person, he at last made all
things clear to me, so that I could now readily distinguish
between a Circle and a Sphere, a Plane Figure and a Solid.

55   This was the Climax, the Paradise, of my strange eventful
History. Henceforth I have to relate the story of my miserable Fall: – most miserable, yet surely most undeserved! For
why should the thirst for knowledge be aroused, only to be
disappointed and punished! My volition shrinks from the
60   painful task of recalling my humiliation; yet, like a second
Prometheus, I will endure this and worse, if by any means I
may arouse in the interiors of Plane and Solid Humanity a
spirit of rebellion against the Conceit which would limit our
Dimensions to Two or Three or any number short of Infinity.
65   Away then with all personal considerations! Let me continue
to the end, as I began, without further digressions or anticipations, pursuing the plain path of dispassionate History.
The exact facts, the exact words, – and they are burnt in upon
my brain, – shall be set down without alteration of an iota;
70   and let my Readers judge between me and Destiny.

19.56. strange eventful History. A phrase taken from the "Seven ages of man speech" in Shakespeare's *As You Like It* 2.7: "Last scene of all / That ends this strange eventful history / Is second childishness and mere oblivion."

19.57. miserable Fall. An allusion to the fall of man, the sudden lapse into a sinful state produced by the act of disobedience of Adam and Eve.

19.61. Prometheus. "[T]he religious element (in Greek literature) is perhaps at its highest in the Prometheus Bound, where a human hero appears to be protesting against the injustice of the gods" (Abbott 1897, 95). There are several variants of the myth of the demigod Prometheus; the following is a brief summary of Aeschylus's *Prometheus Bound*, which represents Prometheus as a martyr in the cause of man.

Shortly after coming to power, Zeus resolved to replace the existing human race with another, but Prometheus thwarted Zeus's plan by stealing fire from Olympus and bringing it to mortals. For this beneficence to mankind, Zeus sentenced Prometheus to a terrible punishment. The play opens as Prometheus is being chained to a rock in Scythia, where he will be subjected to unending torture. Despite his suffering and the urging of persons who come and speak to him, Prometheus refuses to submit to the authority of Zeus. With his last words, he protests, "Behold me, I am wronged" (Hamilton 1937).

19.63. Conceit. Here "conceit" could plausibly mean either idea, especially a fanciful idea, or an excessive appreciation of one's own opinion.

19.65. Away with. Let us be rid of; take away.

19.69. without alteration of an iota. Iota is the smallest letter in the Greek alphabet; hence the figurative meaning, a very small quantity. Curiously, the word "alteration" contains an alteration of "an iota."

19.70. Destiny. The mythological goddess who determines the course of human life.

The Sphere would willingly have continued his lessons by indoctrinating me in the conformation of all regular Solids, Cylinders, Cones, Pyramids, Pentahedrons, Hexahedrons, Dodecahedrons, and Spheres: but I ventured to interrupt
75   him. Not that I was wearied of knowledge. On the contrary, I thirsted for yet deeper and fuller draughts than he was offering to me.

"Pardon me," said I, "O Thou Whom I must no longer address as the Perfection of all Beauty; but let me beg thee
80   to vouchsafe thy servant a sight of thine interior."

Sphere. "My what?"

I. "Thine interior: thy stomach, thy intestines."

Sphere. "Whence this ill-timed impertinent request? And what mean you by saying that I am no longer the Perfection
85   of all Beauty?"

I. My Lord, your own wisdom has taught me to aspire to One even more great, more beautiful, and more closely approximate to Perfection than yourself. As you yourself, superior to all Flatland forms, combine many Circles in
90   One, so doubtless there is One above you who combines many Spheres in One Supreme Existence, surpassing even the Solids of Spaceland. And even as we, who are now in Space, look down on Flatland and see the insides of all things, so of a certainty there is yet above us some higher, purer
95   region, whither thou dost surely purpose to lead me – O Thou Whom I shall always call, everywhere and in all Dimensions, my Priest, Philosopher, and Friend – some yet more spacious Space, some more dimensionable Dimensionality, from the

19.72. regular Solids. The three-dimensional analogue of a regular polygon is a regular solid (or regular polyhedron), a convex solid with plane faces where each face is the same regular polygon and all the faces are arranged in exactly the same way at each vertex (corner). There are only five regular solids in three-dimensional space: a tetrahedron (four triangular faces and four vertices), a cube (six square faces and eight vertices), an octahedron (eight triangular faces and six vertices), a dodecahedron (twelve pentagonal faces and twenty vertices), and an icosahedron (twenty triangular faces and twelve vertices).

Figure 19.2. The regular polyhedra (tetrahedron, cube, octahedron, dodecahedron, icosahedron).

Euclid concludes the *Elements* with a proof that these five polyhedra are the only regular ones. Flatlanders could understand the essence of Euclid's argument: There are at most five ways to fit copies of the same regular polygon around a point in their flat space. Although they would be unable to imagine what it means to fold such a configuration up into three-dimensional space, they could still appreciate that there are at most five regular polyhedra in three-dimensional space (Banchoff 1990a, 93–94).

The regular polyhedra are often called the Platonic solids because in his myth of the creation (the *Timaeus*), Plato presents a theory of the structure of matter in which he associates each of the classical elements (fire, air, water, and earth) with a regular polyhedron (respectively, tetrahedron, octahedron, icosahedron, and cube). Concerning the dodecahedron, Plato says that there remained a fifth construction, which the god used "for the whole universe."

19.76. draught. A quantity of liquor drunk at once.

19.79. the Perfection of all Beauty. In his account of the initiation into the mysteries of love, Socrates says that one comes at last to know "the perfection of beauty" (*Symposium*, 211c–d).

19.95. purpose. Intend.

19.97. Priest, Philosopher, and Friend. Alexander Pope addresses his satirical poem, *An Essay on Man*, to Henry St. John, Lord Bolingbroke, his "guide, philosopher, and friend."

19.98. dimensionable. Capable of being measured; having dimensions. "Dimensionable" is one of eighteen words in the *Oxford English Dictionary* that are illustrated by quotations from *Flatland*.

vantage-ground of which we shall look down together upon
100   the revealed insides of Solid things, and where thine own
intestines, and those of thy kindred Spheres, will lie exposed
to the view of the poor wandering exile from Flatland, to
whom so much has already been vouchsafed.

Sphere. Pooh! Stuff! Enough of this trifling! The time is
105   short, and much remains to be done before you are fit to
proclaim the Gospel of Three Dimensions to your blind
benighted countrymen in Flatland.

I. Nay, gracious Teacher, deny me not what I know it is in
thy power to perform. Grant me but one glimpse of thine
110   interior, and I am satisfied for ever, remaining henceforth
thy docile pupil, thy unemancipable slave, ready to receive
all thy teachings and to feed upon the words that fall from
thy lips.

Sphere. Well, then, to content and silence you, let me say
115   at once, I would show you what you wish if I could; but I
cannot. Would you have me turn my stomach inside out to
oblige you?

I. But my Lord has shown me the intestines of all my countrymen in the Land of Two Dimensions by taking me with
120   him into the Land of Three. What therefore more easy than
now to take his servant on a second journey into the blessed
region of the Fourth Dimension, where I shall look down
with him once more upon this land of Three Dimensions,
and see the inside of every three-dimensioned house, the
125   secrets of the solid earth, the treasures of the mines in Spaceland, and the intestines of every solid living creature, even
of the noble and adorable Spheres.

Sphere. But where is this land of Four Dimensions?

I. I know not: but doubtless my Teacher knows.

130   *Sphere.* Not I. There is no such land. The very idea of it is
utterly inconceivable.

19.110. Grant me but one glimpse of thine interior and I am satisfied for ever. Compare John 14:8: "Philip saith unto him, Lord, show us the Father, and it sufficeth us."

19.111. unemancipable slave. One incapable of being set free.

19.122. the Fourth Dimension. The first person to assert the existence of a fourth dimension of space was the Cambridge Platonist, Henry More, a contemporary of Isaac Newton. More argued that spatial magnitude is a property not only of matter but also of spirits. To allow a spirit to occupy a greater or lesser region of three-dimensional space without changing its "total being," he posits the existence of a fourth dimension, which he calls "Essential Spissitude," from the Latin *spissitudo*, meaning thickness. According to More, what is lost as a Spirit contracts in one or more of its three spatial dimensions is "kept safe in Essential Spissitude" (Cajori 1926, 399–401; MacKinnon 1925, 213).

In *The Kernel and the Husk*, Abbott makes it clear that, unlike Henry More, he does not believe that spirits are "beings of the fourth dimension."

> "You know – or might know if you would read a little book recently published called *Flatland*, and still better, if you would study a very able and original work by Mr. C. H. Hinton – that a being of Four Dimensions, if such there were, could come into our closed rooms without opening door or window, nay, could even penetrate into, and inhabit, our bodies; that he could simultaneously see the insides of all things and the interior of the whole earth thrown open to his vision: he would also have the power of making himself visible and invisible at pleasure; and could address words to us from an invisible position outside us, or inside our own person. Why then might not spirits be beings of the Fourth Dimension? Well, I will tell you why. Although we cannot hope ever to comprehend what a spirit is – just as we can never comprehend what God is – yet St. Paul teaches us that the deep things of the spirit are in some degree made known to us by our own spirits. Now when does the spirit seem most active in us? or when do we seem nearest to the apprehension of 'the deep things of God'? Is it not when we are exercising those virtues which, as St. Paul says, 'abide' – I mean faith, hope and love? Now there is obviously no connection between these virtues and the Fourth Dimension. Even if we could conceive of space of Four Dimensions – which we cannot do, although we can perhaps describe what some of its phenomena would be if it existed – we should not be a whit the better morally or spiritually. It seems to me rather a moral than an intellectual process, to approximate to the conception of a spirit: and toward this no knowledge of Quadridimensional space can guide us" (Abbott 1886, 259).

19.127. adorable. Worthy of worship, not the increasingly trivial current use (charming, delightful).

19.130. There is no such land. More than anything else, the visitation by a being from another world distinguishes *Flatland* from the parable of the cave as well as all other "dimensional stories." The Sphere, whom the Square originally calls "the Stranger," is strange indeed. Although he is supernatural, he is a flawed being. [continued]

I. Not inconceivable, my Lord, to me, and therefore still less inconceivable to my Master. Nay, I despair not that, even here, in this region of Three Dimensions, your Lordship's art
135   may make the Fourth Dimension visible to me; just as in the Land of Two Dimensions my Teacher's skill would fain have opened the eyes of his blind servant to the invisible presence of a Third Dimension, though I saw it not.

Let me recall the past. Was I not taught below that when
140   I saw a Line and inferred a Plane, I in reality saw a Third unrecognized Dimension, not the same as brightness, called "height"? And does it not now follow that, in this region, when I see a Plane and infer a Solid, I really see a Fourth unrecognized Dimension, not the same as colour, but exis-
145   tent, though infinitesimal and incapable of measurement?

And besides this, there is the Argument from Analogy of Figures.

Sphere. Analogy! Nonsense: what analogy?

I. Your Lordship tempts his servant to see whether he
150   remembers the revelations imparted to him. Trifle not with me, my Lord; I crave, I thirst, for more knowledge. Doubtless we cannot *see* that other higher Spaceland now, because we have no eye in our stomachs. But, just as there *was* the realm of Flatland, though that poor puny Lineland Monarch could
155   neither turn to left nor right to discern it, and just as there *was* close at hand, and touching my frame, the land of Three Dimensions, though I, blind senseless wretch, had no power to touch it, no eye in my interior to discern it, so of a surety there is a Fourth Dimension, which my Lord perceives with
160   the inner eye of thought. And that it must exist my Lord himself has taught me. Or can he have forgotten what he himself imparted to his servant?

In One Dimension, did not a moving Point produce a Line with *two* terminal points?

165   In Two Dimensions, did not a moving Line produce a Square with *four* terminal points?

19.131. inconceivable. Nearly all nineteenth-century mathematicians agreed with Samuel Roberts, the president of the London Mathematical Society, who maintained, "It is admitted, on all hands, that we can form no conception whatever of a fourth geometrical dimension" (Roberts 1882, 12). At the same time, many of these men agreed with J. J. Sylvester, who asserted that the inconceivability of four-dimensional space was not a valid reason to deny it any meaning or to exclude it from study.

Sylvester's argument was threefold: (1) He and others had "given evidence of the practical utility of handling space of four dimensions, as if it were conceivable space." (2) The properties of four-dimensional objects "admit of being studied to a great extent, if not completely," by their projections into three-dimensional space. (3) "In philosophy, as in aesthetic, the highest knowledge comes by faith," and such luminaries as Gauss, Cayley, Riemann, and Clifford "have an inner assurance of the reality of transcendental (four-dimensional) space" (Sylvester 1869, 238). For the nineteenth-century British view of the nature of space, see Richards (1988, 54–59).

Abbott would strongly endorse Sylvester's third point. In *The Kernel and the Husk*, he compares faith and imagination in plane geometry and in religion. In Part II of *Flatland*, he has altered fundamentally the nature of this comparison: The hypothetical geometer in *The Kernel and the Husk*, who believes in "a perfect circle by Faith," has seen a material approximation of such a figure (Abbott 1886, 32). By contrast, the Square has no comparable "material reality" that would enable him to imagine a three-dimensional object, and his conjecture that four-dimensional space exists is based entirely on faith.

19.132. Not inconceivable. Lines 19.132 through 19.148 were added to the first edition.

19.136. fain. Eagerly, willingly.

19.144. And does it not ... a Fourth unrecognized Dimension. This fallacious argument, which is analogous to the one added to Section 16, is the most significant of the fourteen lines added to Section 19 (see Appendix A2, Footnote 2).

19.160. inner eye of thought. Imagination.

In Three Dimensions, did not a moving Square produce – did not this eye of mine behold it – that blessed Being, a Cube, with *eight* terminal points?

170   And in Four Dimensions shall not a moving Cube – alas, for Analogy, and alas for the Progress of Truth, if it be not so – shall not, I say, the motion of a divine Cube result in a still more divine Organization with *sixteen* terminal points?

Behold the infallible confirmation of the Series, 2, 4, 8,
175   16: is not this a Geometrical Progression? Is not this – if I might quote my Lord's own words – "strictly according to Analogy"?

Again, was I not taught by my Lord that as in a Line there are *two* bounding Points, and in a Square there are *four*
180   bounding Lines, so in a Cube there must be *six* bounding Squares? Behold once more the confirming Series, 2, 4, 6: is not this an Arithmetical Progression? And consequently does it not of necessity follow that the more divine offspring of the divine Cube in the Land of Four Dimensions, must
185   have 8 bounding Cubes: and is not this also, as my Lord has taught me to believe, "strictly according to Analogy"?

O, my Lord, my Lord, behold, I cast myself in faith upon conjecture, not knowing the facts; and I appeal to your Lord-ship to confirm or deny my logical anticipations. If I am
190   wrong, I yield, and will no longer demand a Fourth Dimen-sion; but, if I am right, my Lord will listen to reason.

I ask therefore, is it, or is it not, the fact, that ere now your countrymen also have witnessed the descent of Beings of a higher order than their own, entering closed rooms, even as
195   your Lordship entered mine, without the opening of doors or windows, and appearing and vanishing at will? On the reply to this question I am ready to stake everything. Deny it, and I am henceforth silent. Only vouchsafe an answer.

Sphere (after a pause). It is reported so. But men are divided
200   in opinion as to the facts. And even granting the facts, they explain them in different ways. And in any case, however great may be the number of different explanations, no one has adopted or suggested the theory of a Fourth Dimension.

19.188. I cast myself in faith upon conjecture. The Square realizes that his analogical argument does not establish the existence of four-dimensional space; his belief that it exists is a "leap of faith." He has used a clever expression – a conjecture is literally a throwing or casting together.

19.196. entering closed rooms . . . without the opening of doors or windows. John records that on two occasions Jesus suddenly appeared in a room despite "the doors being shut" (John 20:19, 26).

19.203. no one has adopted or suggested the theory of the Fourth Dimension. The mathematician William A. Granville makes just this suggestion. Referring to the passages in John 20, he says, "Christ, considered as a higher-dimensional being, certainly had the power to appear in his body as described above, or to do anything else which cannot be done by us in our space of three dimensions but which is possible in our hypothetical space of four dimensions by those who may dwell there" (Granville 1922, 52–53).

Therefore, pray have done with this trifling, and let us return
205   to business.

I. I was certain of it. I was certain that my anticipations
would be fulfilled. And now have patience with me and
answer me yet one more question, best of Teachers! Those
who have thus appeared – no one knows whence – and have
210   returned – no one knows whither – have they also contracted
their sections and vanished somehow into that more Spacious Space, whither I now entreat you to conduct me?

Sphere (*moodily*). They have vanished, certainly – if they
ever appeared. But most people say that these visions arose
215   from the thought – you will not understand me – from the
brain; from the perturbed angularity of the Seer.

I. Say they so? Oh, believe them not. Or if it indeed be
so, that this other Space is really Thoughtland, then take
me to that blessed Region where I in Thought shall see the
220   insides of all solid things. There, before my ravished eye, a
Cube, moving in some altogether new direction, but strictly
according to Analogy, so as to make every particle of his
interior pass through a new kind of Space, with a wake of
its own – shall create a still more perfect perfection than him-
225   self, with sixteen terminal Extra-solid angles, and Eight solid
Cubes for his Perimeter. And once there, shall we stay our
upward course? In that blessed region of Four Dimensions,
shall we linger on the threshold of the Fifth, and not enter
therein? Ah, no! Let us rather resolve that our ambition shall
230   soar with our corporal ascent. Then, yielding to our intellectual onset, the gates of the Sixth Dimension shall fly open;
after that a Seventh, and then an Eighth –

How long I should have continued I know not. In vain
did the Sphere, in his voice of thunder, reiterate his com-
235   mands of silence, and threaten me with the direst penalties if
I persisted. Nothing could stem the flood of my ecstatic aspi-
rations. Perhaps I was to blame; but indeed I was intoxicated
with the recent draughts of Truth to which he himself had
introduced me. However, the end was not long in coming.

19.204. have done with. Stop, cease.

19.216. perturbed angularity of the Seer. Geometrically, having perturbed angularity means having a slightly altered angle; figuratively, as here, it means mentally disturbed.

19.220. ravished. Transported with ecstasy or delight. One meaning of "to ravish" is to remove from one place or state to another, as from Earth to Heaven.

19.221. a Cube, moving. The Square is describing what is now called a hypercube, the four-dimensional analogue of a cube, which is formed when a cube moves perpendicular to itself in four-dimensional space. For more about hypercubes, see Banchoff (1990a) and Rucker (1984).

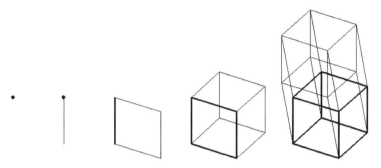

Figure 19.3. The derivation sequence extended to the fourth dimension.

19.224. every particle ... with a wake of its own. The derivation sequence is a refinement of an older theory of the generation of geometric figures from numbers according to which *one* is associated with the point, *two* with the line segment, *three* with the triangle, and *four* with the pyramid (tetrahedron).

We may obtain the sequence, point-line-triangle-tetrahedron, by a generalization of the derivation sequence in which each point moves separately (in the Square's words, "with a wake of its own"). A point moves in a constant direction to form a line segment; each point of the line segment moves in a direction perpendicular to the line segment to form a triangle; each point of the triangle moves in a direction perpendicular to the triangle to form a tetrahedron.

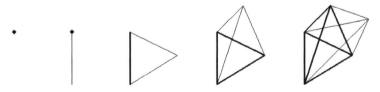

Figure 19.4. Point, line, triangle, tetrahedron, hypertetrahedron (5-cell).

The medieval Scholastic philosopher Nicole Oresme described the essence of such a construction in his *Treatise on the configurations of qualities and motions* (c. 1355). (For Oresme, such figures are not actually constructed but "imagined.") [continued]

240 My words were cut short by a crash outside, and a simultaneous crash inside me, which impelled me through Space
with a velocity that precluded speech. Down! down! down!
I was rapidly descending; and I knew that return to Flatland was my doom. One glimpse, one last and never-to-be-
245 forgotten glimpse I had of that dull level wilderness – which
was now to become my Universe again – spread out before
my eye. Then a darkness. Then a final, all-consummating
thunder-peal; and, when I came to myself, I was once more
a common creeping Square, in my Study at home, listening
250 to the Peace-Cry of my approaching Wife.

§20
How the Sphere encouraged me in a Vision

ALTHOUGH I HAD LESS THAN a minute for reflection, I felt,
by a kind of instinct, that I must conceal my experiences
from my Wife. Not that I apprehended, at the moment, any
danger from her divulging my secret, but I knew that to
5 any Woman in Flatland the narrative of my adventures must
needs be unintelligible. So I endeavoured to reassure her by
some story, invented for the occasion, that I had accidentally
fallen through the trap-door of the cellar, and had there lain
stunned.

10 The Southward attraction in our country is so slight that
even to a Woman my tale necessarily appeared extraordinary
and well-nigh incredible; but my Wife, whose good sense far

19.225. Extra-. The prefix "extra-" conveys the general sense of "lying outside the province or scope of something" – in this case, three-dimensional space. The current prefix meaning "the analogue in a space of four or more dimensions" is "hyper," which was first used in this sense by J. J. Sylvester: hyperlocus (1851); hyperplane, hyperpyramid, and hypergeometry (1863) (Manning 1914, 329).

19.226. Eight solid Cubes. The table gives the number of k-cubes in an n-cube; the entries in row n are the coefficients of the expansion of the binomial $(2x + y)^n$. For example, the numbers in row 3 are the coefficients of $8x^3 + 12x^2y + 6xy^2 + y^3$.

n	n-cube	0-cubes	1-cubes	2-cubes	3-cubes	4-cubes
0	point	1				
1	segment	2	1			
2	square	4	4	1		
3	cube	8	12	6	1	
4	hypercube	16	32	24	8	1

19.238. intoxicated with the recent draughts of Truth. In "An essay on criticism," Alexander Pope develops the metaphor of learning as drinking with the paradox that learning (drinking) in small amounts is intoxicating but learning (drinking) a lot is sobering.

> A *little learning* is a dangerous thing;
> Drink deep, or taste not the Pierian spring:
> There shallow draughts intoxicate the brain,
> And drinking largely sobers us again.

Abbott was well acquainted with Pope's poetry; he contributed a substantial introduction to his father's *A Concordance to the Works of Alexander Pope* (1875).

19.245. wilderness. In religious uses, "wilderness" often refers to the present world or life as contrasted with Heaven or the future life.

19.248. all-consummating thunder-peal. An allusion to the final scene of *Prometheus Bound*, where thunder peals as the crumbling universe falls upon Prometheus.

19.249. creeping Square. An allusion to the serpent in the Garden of Eden. In describing his fate, the Square several times alludes to the biblical account of Adam's fall and to its closest Greek equivalent, the myth of Prometheus. Although there are distinctions between these two stories, both portray imagination as an offense against the gods (Kearney 1988, 79–87). *Flatland* also tells a story of "fallen imagination," for it is the Square's offending the Sphere by imagining the possibility of a fourth dimension that is the proximate cause of his "miserable Fall."

exceeds that of the average of her Sex, and who perceived that I was unusually excited, did not argue with me on the
15   subject, but insisted that I was ill and required repose. I was glad of an excuse for retiring to my chamber to think quietly over what had happened. When I was at last by myself, a drowsy sensation fell on me; but before my eyes closed I endeavoured to reproduce the Third Dimension,
20   and especially the process by which a Cube is constructed through the motion of a Square. It was not so clear as I could have wished; but I remembered that it must be "Upward, and yet not Northward," and I determined steadfastly to retain these words as the clue which, if firmly grasped, could not
25   fail to guide me to the solution. So mechanically repeating, like a charm, the words, "Upward yet not Northward," I fell into a sound refreshing sleep.

During my slumber I had a dream. I thought I was once more by the side of the Sphere, whose lustrous hue beto-
30   kened that he had exchanged his wrath against me for perfect placability. We were moving together towards a bright but infinitesimally small Point, to which my Master directed my attention. As we approached, methought there issued from it a slight humming noise as from one of your Spaceland
35   blue-bottles, only less resonant by far, so slight indeed that even in the perfect stillness of the Vacuum through which we soared, the sound reached not our ears till we checked our flight at a distance from it of something under twenty human diagonals.

40   "Look yonder," said my Guide, "in Flatland thou hast lived; of Lineland thou hast received a vision; thou hast soared with me to the heights of Spaceland; now, in order to complete the range of thy experience, I conduct thee down-ward to the lowest depth of existence, even to the realm of
45   Pointland, the Abyss of No Dimensions.

Notes on Section 20.

20.13. good sense far exceeds. The Square recognizes that his wife has considerable intelligence despite her lack of education, yet he does not seem to have entertained the possibility that the same thing is true of other women.

20.28. I had a dream. At line 16.160, the Sphere revealed that he can discern the Square's "phantasmal visions," and he may even have the power to cause the Square to have dreams or visions.

20.35. blue-bottles. Flies (so called because of their blue bodies); the adult is nearly twice the size of a housefly.

20.37. Vacuum ... sound. Sound cannot travel in a vacuum (i.e., a space entirely empty of matter).

20.39. twenty human diagonals. Twenty feet.

20.45. No Dimensions/non-dimensional. The dimension of a space may be defined informally as the maximum number of independent motions available for movement in the space. Because movement in a one-point space is impossible, such a space has dimension zero; in the Square's words, it has no dimensions, or is non-dimensional.

20.45. Abyss/Gulf. A deep hollow or chasm.

"Behold yon miserable creature. That Point is a Being like
ourselves, but confined to the non-dimensional Gulf. He is
himself his own World, his own Universe; of any other than
himself he can form no conception; he knows not Length, nor
50   Breadth, nor Height, for he has had no experience of them;
he has no cognizance even of the number Two; nor has he a
thought of Plurality; for he is himself his One and All, being
really Nothing. Yet mark his perfect self-contentment, and
hence learn this lesson, that to be self-contented is to be vile
55   and ignorant, and that to aspire is better than to be blindly
and impotently happy. Now listen."

He ceased; and there arose from the little buzzing creature
a tiny, low, monotonous, but distinct tinkling, as from one
of your Spaceland phonographs, from which I caught these
60   words, "Infinite beatitude of existence! It is; and there is none
else beside It."

"What," said I, "does the puny creature mean by 'it'?"
"He means himself," said the Sphere: "have you not noticed
before now, that babies and babyish people who cannot dis-
65   tinguish themselves from the world, speak of themselves in
the Third Person? But hush!"

"It fills all Space," continued the little soliloquizing Creature, "and what It fills, It is. What It thinks, that It utters; and
what It utters, that It hears; and It itself is Thinker, Utterer,
70   Hearer, Thought, Word, Audition; it is the One, and yet the
All in All. Ah, the happiness ah, the happiness of Being!"

"Can you not startle the little thing out of its complacency?" said I. "Tell it what it really is, as you told me; reveal
to it the narrow limitations of Pointland, and lead it up to
75   something higher." "That is no easy task," said my Master;
"try you."

Hereon, raising my voice to the uttermost, I addressed the
Point as follows:

20.47. a Being like ourselves. As Socrates begins telling the cave parable, Glaucon remarks that it is a strange image and they are strange prisoners, and Socrates replies that the prisoners are like us (*Republic*, 515a).

20.49. of any other than himself he can form no conception. The Point is a solipsist – he cannot attach any meaning to the idea that there could be thoughts, experiences, or emotions other than his own.

20.59. tinkling, as from one of your Spaceland phonographs. In Thomas Edison's first phonograph recordings, sound was recorded as indentations embossed into a sheet of tinfoil that was wrapped around a cylinder. *The Times* reported that Edison's phonograph reproduced the words of a speaker "in his own voice, tinged, however, with a slight metallic or mechanical tone" (*The Times*, 17 January 1878, 4).

20.60. beatitude. Supreme blessedness or happiness; perfection. The Beatitudes (Matthew 5:3–11 and Luke 6:20–22) describe the qualities of Christian perfection.

80 "Silence, silence, contemptible Creature. You call yourself the All in All, but you are the Nothing: your so-called Universe is a mere speck in a Line, and a Line is a mere shadow as compared with – " "Hush, hush, you have said enough," interrupted the Sphere, "now listen, and mark the effect of your harangue on the King of Pointland."

85 The lustre of the Monarch, who beamed more brightly than ever upon hearing my words, showed clearly that he retained his complacency; and I had hardly ceased when he took up his strain again. "Ah, the joy, ah, the joy of Thought! What can It not achieve by thinking! Its own Thought coming 90 to Itself, suggestive of Its disparagement, thereby to enhance Its happiness! Sweet rebellion stirred up to result in triumph! Ah, the divine creative power of the All in One! Ah, the joy, the joy of Being!"

"You see," said my Teacher, "how little your words have 95 done. So far as the Monarch understands them at all, he accepts them as his own – for he cannot conceive of any other except himself – and plumes himself upon the variety of 'Its Thought' as an instance of creative Power. Let us leave this God of Pointland to the ignorant fruition of his omnipresence 100 and omniscience: nothing that you or I can do can rescue him from his self-satisfaction."

After this, as we floated gently back to Flatland, I could hear the mild voice of my Companion pointing the moral of my vision, and stimulating me to aspire, and to teach others 105 to aspire. He had been angered at first – he confessed – by my ambition to soar to Dimensions above the Third; but, since then, he had received fresh insight, and he was not too proud to acknowledge his error to a Pupil. Then he proceeded to initiate me into mysteries yet higher than those I had wit- 110 nessed, showing me how to construct Extra-Solids by the motion of Solids, and Double Extra-Solids by the motion of Extra-Solids, and all "strictly according to Analogy," all by methods so simple, so easy, as to be patent even to the Female Sex.

20.99. fruition. Enjoyment.

20.103. point the moral. To explain the moral meaning of.

20.109. initiate me into mysteries. Abbott emphasizes that the Square's introduction to higher dimensions is a figure for a revelation of higher truth by referring to it several times as "his initiation into the mysteries." In theological usage, a "mystery" means a religious truth known only from divine revelation, but Abbott's second-order metaphor refers to something even more specific. In ancient Greece, "mysteries" were secret rituals that aimed at changing the mind of the initiate through an experience of the sacred (Burkert 1987, 11). There were many "mystery cults" among the Greeks; the earliest and most influential was the one at Eleusis.

Plato uses the language of mysteries in a number of dialogues as an image of intellectual or moral transformation; two important examples are the poetic metaphors in the *Symposium* and *Phaedrus*. There is no explicit mention of the mysteries in the parable of the cave, but the emergence from the cave into the sunlight parallels the culmination of the ritual journey in which a sacred vision was suddenly revealed to the initiate.

20.110. Extra-Solids. An "extra-solid" is an object in four-dimensional space; what the Square calls an Extra-Cube is now called a hypercube. C. Howard Hinton first called the four-dimensional analogue of a cube a "four-square"; later, he settled on the name "tessaract" (now spelled "tesseract") (Hinton 1880; 1888). Higher-dimensional polyhedra are now called "polytopes," and the standard notation for a four-dimensional, regular polytope is "k-cell," where k indicates the number of three-dimensional boundary cells. Thus, a hypercube is an 8-cell.

In two-dimensional space, there are n-sided regular polygons for every integer n greater than two. In three-dimensional space, there are just five regular polyhedra (see Note 19.72). In a paper written in 1852 but not published until 1901, the Swiss geometer Ludwig Schläfli determined all of the higher-dimensional regular polytopes. He showed that in four-dimensional space there are just six regular polytopes: the 5-cell, 8-cell, 16-cell, 120-cell, and 600-cell (analogues of the tetrahedron, cube, octahedron, dodecahedron, and icosahedron, respectively) as well as the 24-cell, which has no three-dimensional analogue. Further, he showed that for every n greater than 4, the only regular polytopes are the n-dimensional analogues of the tetrahedron, the cube, and the octahedron (Stillwell 2001, 21–22).

20.111. Double Extra-Solids. A double extra-solid is a five-dimensional object.

§21
How I tried to teach the Theory of Three Dimensions to my Grandson, and with what success

I AWOKE REJOICING, AND BEGAN to reflect on the glorious career before me. I would go forth, methought, at once, and evangelize the whole of Flatland. Even to Women and Soldiers should the Gospel of Three Dimensions be proclaimed.
5 I would begin with my Wife.

Just as I had decided on the plan of my operations, I heard the sound of many voices in the street commanding silence. Then followed a louder voice. It was a herald's proclamation. Listening attentively, I recognized the words of the Resolu-
10 tion of the Council, enjoining the arrest, imprisonment, or execution of any one who should pervert the minds of the people by delusions, and by professing to have received revelations from another World.

I reflected. This danger was not to be trifled with. It would
15 be better to avoid it by omitting all mention of my Revelation, and by proceeding on the path of Demonstration – which after all, seemed so simple and so conclusive that nothing would be lost by discarding the former means. "Upward, not Northward" – was the clue to the whole proof. It had
20 seemed to me fairly clear before I fell asleep; and when I first awoke, fresh from my dream, it had appeared as patent as Arithmetic; but somehow it did not seem to me quite so obvious now. Though my Wife entered the room opportunely just at that moment, I decided, after we had interchanged a few
25 words of commonplace conversation, not to begin with her.

Notes on Section 21.

21.16. Demonstration. According to Aristotle, a demonstration (*apodeixis*) is a deduction that produces knowledge: "By demonstration I mean a scientific deduction; and by scientific I mean one in virtue of which, by having it, we understand something" (*Posterior Analytics*, 70b).

My Pentagonal Sons were men of character and standing, and physicians of no mean reputation, but not great in
mathematics, and, in that respect, unfit for my purpose. But
it occurred to me that a young and docile Hexagon, with
30   a mathematical turn, would be a most suitable pupil. Why
therefore not make my first experiment with my little pre-
cocious Grandson, whose casual remarks on the meaning of
3^3 had met with the approval of the Sphere? Discussing the
matter with him, a mere boy, I should be in perfect safety; for
35   he would know nothing of the Proclamation of the Council;
whereas I could not feel sure that my Sons – so greatly did
their patriotism and reverence for the Circles predominate
over mere blind affection – might not feel compelled to hand
me over to the Prefect, if they found me seriously maintain-
40   ing the seditious heresy of the Third Dimension.

But the first thing to be done was to satisfy in some way
the curiosity of my Wife, who naturally wished to know
something of the reasons for which the Circle had desired
that mysterious interview, and of the means by which he had
45   entered the house. Without entering into the details of the
elaborate account I gave her, – an account, I fear, not quite
so consistent with truth as my Readers in Spaceland might
desire, – I must be content with saying that I succeeded at
last in persuading her to return quietly to her household
50   duties without eliciting from me any reference to the World
of Three Dimensions. This done, I immediately sent for my
Grandson; for, to confess the truth, I felt that all that I had
seen and heard was in some strange way slipping away from
me, like the image of a half-grasped, tantalizing dream, and
55   I longed to essay my skill in making a first disciple.

When my Grandson entered the room I carefully secured
the door. Then, sitting down by his side and taking our mathematical tablets, – or, as you would call them, Lines – I told
him we would resume the lesson of yesterday. I taught him
60   once more how a Point by motion in One Dimension produces a Line, and how a straight Line in Two Dimensions
produces a Square. After this, forcing a laugh, I said, "And
now, you scamp, you wanted to make me believe that a

Square may in the same way by motion 'Upward, not North-
65   ward' produce another figure, a sort of extra Square in Three
Dimensions. Say that again, you young rascal."

At this moment we heard once more the herald's "O yes!
O yes!" outside in the street proclaiming the Resolution of
the Council. Young though he was, my Grandson – who was
70   unusually intelligent for his age, and bred up in perfect reverence for the authority of the Circles – took in the situation
with an acuteness for which I was quite unprepared. He
remained silent till the last words of the Proclamation had
died away, and then, bursting into tears, "Dear Grandpapa,"
75   he said, "that was only my fun, and of course I meant nothing at all by it; and we did not know anything then about the
new Law; and I don't think I said anything about the Third
Dimension; and I am sure I did not say one word about
'Upward, not Northward,' for that would be such nonsense,
80   you know. How could a thing move Upward, and not Northward? Upward and not Northward! Even if I were a baby, I
could not be so absurd as that. How silly it is! Ha! ha! ha!"

"Not at all silly," said I, losing my temper; "here for example, I take this Square," and, at the word, I grasped a move-
85   able Square, which was lying at hand – "and I move it, you
see, not Northward but – yes, I move it Upward – that is to
say, not Northward, but I move it somewhere – not exactly
like this, but somehow – " Here I brought my sentence to an
inane conclusion, shaking the Square about in a purposeless
90   manner, much to the amusement of my Grandson, who burst
out laughing louder than ever, and declared that I was not
teaching him, but joking with him. So saying he unlocked
the door and ran out of the room; and thus ended my first
attempt to convert a pupil to the Gospel of Three Dimensions.

21.68 herald's "O yes! O yes!" In Athens, heralds (*kēryx*) were attached to various officials and government boards. Among various responsibilities, the herald summoned the Council and the Assembly and (as here) made state proclamations.

The use of public criers in the British Isles dates back to Norman times, when the cry "oyez, oyez, oyez" (old French for "hear ye") was used to command the silence and attention of the (mostly illiterate) public to a proclamation, which was about to be read.

21.72. acuteness. The boy was quick to understand that his earlier questions about the meaning of 3^3 in geometry were related to the herald's pronouncement, and he disowned them at once.

§22
How I then tried to diffuse the Theory of Three Dimensions by other means, and of the result

MY FAILURE WITH MY GRANDSON did not encourage me to communicate my secret to others of my household; yet neither was I led by it to despair of success. Only I saw that I must not wholly rely on the catch-phrase, "Upward, not
5   Northward," but must rather endeavour to seek a demonstration by setting before the public a clear view of the whole subject; and for this purpose it seemed necessary to resort to writing.

So I devoted several months in privacy to the composition
10   of a treatise on the mysteries of Three Dimensions. Only, with the view of evading the Law, if possible, I spoke not of a physical Dimension, but of a Thoughtland whence, in theory, a Figure could look down upon Flatland and see simultaneously the insides of all things, and where it was possible
15   that there might be supposed to exist a Figure environed, as it were, with six Squares, and containing eight terminal Points. But in writing this book I found myself sadly hampered by the impossibility of drawing such diagrams as were necessary for my purpose; for of course, in our country of Flatland,
20   there are no tablets but Lines, and no diagrams but Lines, all in one straight Line and only distinguishable by difference of size and brightness; so that, when I had finished my treatise (which I entitled, "Through Flatland to Thoughtland") I could not feel certain that many would understand my
25   meaning.

Meanwhile my life was under a cloud. All pleasures palled upon me; all sights tantalized and tempted me to outspoken

Notes on Section 22.

22.8. resort to writing. The Square has set himself an impossible task – putting his revelation into words. His words must fail because no other Flatlander has seen what he has seen, and the lack of common experience renders his language meaningless to his countrymen.

The mysteries were *arrheta* (unspeakable), meaning not only that they were to be kept secret from the uninitiated but also that their essence was incapable of being expressed in words (Burkert 1987, 69). They are an ideal figure for the Square's intellectual passage and his subsequent inability to describe his experience to others.

22.16. a Figure environed, as it were, with six Squares. A cube. The *Oxford English Dictionary* defines "environed" as set round with other objects, and conjectures that the word was never in actual English use.

22.17. book. A Flatland book might look like a thread with writing similar to Morse Code. It could be stored by wrapping it around a disc.

22.18. such diagrams as were necessary. According to Reviel Netz, the Greeks regarded diagrams not as appendages to propositions but as the core of propositions (Netz 1999, 35). In modern mathematics, a diagram may serve as a useful adjunct to the proof of a proposition, but it is not a necessary part of the proof.

22.23 "Through Flatland to Thoughtland." Jonathan Smith was the first person to point out that "Through Flatland to Thoughtland" is an allusion to *Through Nature to Christ*, the controversial book in which Abbott first professed the liberal theology that characterized all his subsequent work (Smith 1994, 265). The Square's attempt to avoid arrest by speaking of "Thoughtland" rather than three-dimensional space has a parallel in Abbott's revisions of the manuscript of *Through Nature to Christ*. He was fearful that the publication of this book would lead to his dismissal as headmaster, and he removed its most controversial chapter. He closes a letter to his publisher in which he expresses his apprehension by saying, "I have struck out almost all the 'Dreams' and 'Visions'" (Abbott 1877c). The deleted chapter was published thirty years later as "Revelation by visions and voices" (Abbott 1907b).

22.23. Thoughtland. Abbott was fundamentally a Platonist; he believed in the existence of a realm apprehensible only by the intellect that is distinct from the world perceived by our senses. Further, the former realm is the ultimate origin of the existence and meaning of everything in the latter. In *Apologia*, he urges his readers to try to conceive of the existence of a Thoughtland that is "as much more real than Factland as the land of three dimensions seems to us more real than the land of two" (Abbott 1907a, 83).

22.26. under a cloud. Overshadowed with gloom, or because the Square was "considered heterodox," perhaps he means "under suspicion."

22.27. palled upon me. Became flat and dull.

treason, because I could not but compare what I saw in Two
Dimensions with what it really was if seen in Three, and
30   could hardly refrain from making my comparisons aloud. I
neglected my clients and my own business to give myself to
the contemplation of the mysteries which I had once beheld,
yet which I could impart to no one, and found daily more
difficult to reproduce even before my own mental vision.

35   One day, about eleven months after my return from Spaceland, I tried to see a Cube with my eye closed, but failed; and
though I succeeded afterwards, I was not then quite certain
(nor have I been ever afterwards) that I had exactly realized
the original. This made me more melancholy than before,
40   and determined me to take some step; yet what, I knew not.
I felt that I would have been willing to sacrifice my life for
the Cause, if thereby I could have produced conviction. But
if I could not convince my Grandson, how could I convince
the highest and most developed Circles in the land?

45   And yet at times my spirit was too strong for me, and I
gave vent to dangerous utterances. Already I was considered
heterodox if not treasonable, and I was keenly alive to the
dangers of my position; nevertheless I could not at times
refrain from bursting out into suspicious or half-seditious
50   utterances, even among the highest Polygonal and Circular
society. When, for example, the question arose about the
treatment of those lunatics who said that they had received
the power of seeing the insides of things, I would quote
the saying of an ancient Circle, who declared that prophets
55   and inspired people are always considered by the majority
to be mad; and I could not help occasionally dropping
such expressions as "the eye that discerns the interiors of
things," and "the all-seeing land:" once or twice I even let
fall the forbidden terms "the Third and Fourth Dimensions."
60   At last, to complete a series of minor indiscretions, at a
meeting of our Local Speculative Society held at the palace
of the Prefect himself, – some extremely silly person having
read an elaborate paper exhibiting the precise reasons
why Providence has limited the number of Dimensions
65   to Two, and why the attribute of omnividence is assigned
to the Supreme alone – I so far forgot myself as to give an
exact account of the whole of my voyage with the Sphere

22.32. contemplation of the mysteries. "Plato... can find no more fitting vehicle for his most transcendent thoughts than the imagery which he borrows from the contemplation of the mysteries" (Campbell 1898, 264).

22.56. considered ... to be mad. "According to Plato (*Phaedrus*), philosophers were confused by common people with lunatics, so the visions of disease, which in their extreme forms characterise madness, have been sometimes confused with the visions of faith, which in their extreme form characterise prophets or seers" (Abbott 1907b, 8–9).

22.64. limited the number of Dimensions. For a discussion of arguments for the three-dimensionality of space, see Jammer (1969, 174–186) and Janich (1992).

Carl Friedrich Gauss, one of the greatest of all mathematicians, often said that he regarded the three dimensions of space as a peculiarity of the human mind; people who could not understand this he humorously called "Boeotians." He remarked that just as we can imagine beings that are only aware of two dimensions, so may we conceive of higher (dimensional) beings looking down upon us. He joked that he had set aside certain problems on dimensionality, which he intended to deal with geometrically when he had attained a higher state of existence (Waltershausen 1856, 81).

22.66. forgot myself. Lost sight of the need to be circumspect.

into Space, and to the Assembly Hall in our Metropolis, and then to Space again, and of my return home, and of every-
70   thing that I had seen and heard in fact or vision. At first, indeed, I pretended that I was describing the imaginary experiences of a fictitious person; but my enthusiasm soon forced me to throw off all disguise, and finally, in a fervent peroration, I exhorted all my hearers to divest themselves of
75   prejudice and to become believers in the Third Dimension.

Need I say that I was at once arrested and taken before the Council?

Next morning, standing in the very place where but a very few months ago the Sphere had stood in my company, I was
80   allowed to begin and to continue my narration unquestioned and uninterrupted. But from the first I foresaw my fate; for the President, noting that a guard of the better sort of Policemen was in attendance, of angularity little, if at all, under 55°, ordered them to be relieved before I began my defence,
85   by an inferior class of 2° or 3°. I knew only too well what that meant. I was to be executed or imprisoned, and my story was to be kept secret from the world by the simultaneous destruction of the officials who had heard it; and, this being the case, the President desired to substitute the cheaper for
90   the more expensive victims.

After I had concluded my defence, the President, perhaps perceiving that some of the junior Circles had been moved by my evident earnestness, asked me two questions: –

1. Whether I could indicate the direction which I meant
95   when I used the words "Upward, not Northward"?
2. Whether I could by any diagrams or descriptions (other than the enumeration of imaginary sides and angles) indicate the Figure I was pleased to call a Cube?

I declared that I could say nothing more, and that I must
100   commit myself to the Truth, whose cause would surely prevail in the end.

The President replied that he quite concurred in my sentiment, and that I could not do better. I must be sentenced to perpetual imprisonment; but if the Truth intended that

22.70. seen . . . in fact or vision.   For Abbott, as for the poet William Blake, a "vision" does not have objective reality but exists in the imagination where thoughts comprise "reality." Of Blake, Abbott says, "By quoting none but his wildest and most grotesque utterances, it is quite easy to give the impression that he was a mere madman. . . But his whole life and works prove that he had a faculty of seeing what others cannot see, that which with us is imagination was with him sight" (Abbott 1907b, 16).

22.72. enthusiasm.   From the Greek word *entheos*, meaning "possessed by a god."

22.73. throw off all disguise.   Speak the truth.

22.74. peroration.   The concluding part of a speech or written discourse, in which the speaker or writer gives a forceful summary. Aristotle's *Rhetoric* closes appropriately with a chapter on peroration.

22.104. sentenced to perpetual imprisonment.   Plato warns that if one who has made the journey from the cave into the light should return to the darkness of the cave, he will not be able to explain or describe what he has experienced and will be subjected to ridicule and even persecution (*Republic*, 517a).

105 I should emerge from prison and evangelize the world, the
Truth might be trusted to bring that result to pass. Meanwhile
I should be subjected to no discomfort that was not necessary
to preclude escape, and, unless I forfeited the privilege by
misconduct, I should be occasionally permitted to see my
110 brother who had preceded me to my prison.

Seven years have elapsed and I am still a prisoner, and – if I
except the occasional visits of my brother – debarred from all
companionship save that of my jailers. My brother is one of
the best of Squares, just, sensible, cheerful, and not without
115 fraternal affection; yet I must confess that my weekly inter-
views, at least in one respect, cause me the bitterest pain.
He was present when the Sphere manifested himself in the
Council Chamber; he saw the Sphere's changing sections; he
heard the explanation of the phenomena then given to the
120 Circles. Since that time, scarcely a week has passed during
seven whole years, without his hearing from me a repetition
of the part I played in that manifestation, together with
ample descriptions of all the phenomena in Spaceland, and
the arguments for the existence of Solid things derivable
125 from Analogy. Yet – I take shame to be forced to confess
it – my brother has not yet grasped the nature of the Third
Dimension, and frankly avows his disbelief in the existence
of a Sphere.

Hence I am absolutely destitute of converts, and, for aught
130 that I can see, the millennial Revelation has been made to
me for nothing. Prometheus up in Spaceland was bound
for bringing down fire for mortals, but I – poor Flatland
Prometheus – lie here in prison for bringing down nothing
to my countrymen. Yet I exist in the hope that these mem-
135 oirs, in some manner, I know not how, may find their way
to the minds of humanity in Some Dimension, and may stir
up a race of rebels who shall refuse to be confined to limited
Dimensionality.

That is the hope of my brighter moments. Alas, it is not
140 always so. Heavily weighs on me at times the burdensome
reflection that I cannot honestly say I am confident as to
the exact shape of the once-seen, oft-regretted Cube; and in
my nightly visions the mysterious precept, "Upward, not
Northward," haunts me like a soul-devouring Sphinx. It is

22.111. Seven years. It is significant that the Square has spent seven years in prison since he wrote "Through Flatland to Thoughtland." His failure as an apostle of the Gospel of Three Dimensions parallels Abbott's own frustration in trying to inspire belief in the non-miraculous Christianity that he advocated in *Through Nature to Christ*, published seven years before *Flatland* (see Note 22.23).

22.129. aught. Anything.

22.142. oft-regretted. When the Square speaks of "the oft-regretted cube," he means that he has often thought of the cube with distress and longing – not that he is sorry for something he did or left undone.

22.143. precept. Teaching, especially a religious teaching.

22.144. haunts me like the soul-devouring Sphinx. More properly, the Square might have said, "haunts me like the riddle of the soul-devouring Sphinx."

In his essay, "Sphinx, or science," Francis Bacon describes the Sphinx as a hybrid creature with the face and voice of a virgin, the wings of a bird, and the claws of a griffin. She lived on the top of a mountain near the city of Thebes, where she seized travelers who happened along the highways. Once she had them in her power, she proposed certain difficult riddles. Those captives who could not at once solve and interpret her riddles, she tore to pieces. In Bacon's interpretation, the Sphinx represents science (knowledge), and the riddles she proposes concern the nature of things and the nature of man. Until the riddles are solved, "they strangely torment and worry the mind, pulling it first this way and then that, and fairly tearing it to pieces" (Bacon 1860, 159–162).

145 part of the martyrdom which I endure for the cause of
the Truth that there are seasons of mental weakness, when
Cubes and Spheres flit away into the background of scarce-
possible existences; when the Land of Three Dimensions
seems almost as visionary as the Land of One or None; nay,
150 when even this hard wall that bars me from my freedom,
these very tablets on which I am writing, and all the substantial realities of Flatland itself, appear no better than the
offspring of a diseased imagination, or the baseless fabric of
a dream.

22.145. martyrdom. The sufferings of a martyr, a person who bears witness for a belief (Abbott 1916, xiii).

22.146. the cause of the Truth. "His pupils carried away most enduringly from his teaching a deep impression of an overmastering intellectual honesty and of the ruthless application of all available means to the discovery of truth. . . A distinguished contemporary headmaster. . . declared, 'I never met a man with so strong a passion for truth as Edwin Abbott.' Abbott's greatness as a teacher, preacher, and scholar was based on deep and lively human sympathies and an unquenchable passion for truth" (Obituary 1926a).

22.146. seasons. Periods, spells.

22.149. visionary. Unreal.

22.152. realities. Abbott believes that we know nothing at all about what is real: "As Solidland may be called more 'real' than Flatland, so Thoughtland may be found more real than Factland." He even confesses that he fancies the possibility that there is no such thing as matter but only laws of force (Abbott 1907a, 11, 63).

22.153. diseased imagination. The concept of a "diseased imagination" was well established by the early eighteenth century, and many works were devoted to suggested cures for the malady, now called a neurosis.

22.154. the baseless fabric. Abbott closes Part II as he opened it, with words from Shakespeare's *The Tempest*. The last sentence and the drawing are reminiscent of the famous soliloquy that follows the wedding masque in Act 4 of *The Tempest*. There Prospero declares that what we call reality is, like the "insubstantial pageant" he has just staged for the newlyweds, merely an illusion:

> Our revels now are ended. These our actors,
> As I foretold you, were all spirits and
> Are melted into air, into thin air;
> And like the baseless fabric of this vision,
> The cloud-capped towers, the gorgeous palaces,
> The solemn temples, the great globe itself,
> Yea, all which it inherit, shall dissolve,
> And, like this insubstantial pageant faded,
> Leave not a rack behind. We are such stuff
> As dreams are made on, and our little life
> Is rounded with a sleep.

Epilogue by the Editor

IF MY POOR FLATLAND FRIEND retained the vigour of mind which he enjoyed when he began to compose these Memoirs, I should not now need to represent him in this Epilogue, in which he desires, firstly, to return his thanks to his read-

5 ers and critics in Spaceland, whose appreciation has, with unexpected celerity, required a second edition of his work; secondly, to apologize for certain errors and misprints (for which, however, he is not entirely responsible); and, thirdly, to explain one or two misconceptions. But he is not the Square

10 he once was. Years of imprisonment, and the still heavier burden of general incredulity and mockery, have combined with the natural decay of old age to erase from his mind many of the thoughts and notions, and much also of the terminology, which he acquired during his short stay in Spaceland.

15 He has, therefore, requested me to reply in his behalf to two special objections, one of an intellectual, the other of a moral nature.

The first objection is, that a Flatlander, seeing a Line, sees something that must be *thick* to the eye as well as *long* to

20 the eye (otherwise it would not be visible, if it had not some thickness); and consequently he ought (it is argued) to acknowledge that his countrymen are not only long and broad, but also (though doubtless in a very slight degree) *thick* or *high*. His objection is plausible, and, to Spacelanders,

25 almost irresistible, so that, I confess, when I first heard it, I knew not what to reply. But my poor old friend's answer appears to me completely to meet it.

"I admit," said he – when I mentioned to him this objection – "I admit the truth of your critic's facts, but I

30 deny his conclusions. It is true that we have really in Flatland a Third unrecognized Dimension called 'height,' just as it is also true that you have really in Spaceland a Fourth

Notes on the Epilogue

Title. This "epilogue" was the Preface to the second edition of *Flatland*. The origin of the Preface was a letter, which the Square sent to *The Athenaeum* in response to a review of *Flatland*. Appendix A contains the review, information on the reviewer (A. J. Butler), and the Square's letter with our annotations, which detail connections between the letter and this section.

E.15. requested me to reply. The Square's admission that he has forgotten many of the ideas and much of the language of his own memoir may be Abbott's way of saying that so much time has passed since he began writing *Flatland* that he finds it inconvenient or difficult to reconstruct the Square's literary style. In any case, this preface/epilogue is stylistically different from the rest of the text, and in the second edition it was set in italic type to emphasize that it was not the work of the Square.

E.18. The first objection. This objection was raised by A. J. Butler, who asserts: "Of course, if our friend the Square and his polygonal relations could see each other edgewise, they must have had some thickness, and need not, therefore, have been so distressed at the doctrine of a third dimension" (see Appendix A2, Footnote 2).

E.30. I admit the truth of your critic's facts, but I deny his conclusions. The Square concedes that Flatlanders have thickness, but he maintains correctly that they need not have it to be visible to one another. The most likely source of Butler's misapprehension is that a two-dimensional figure would disappear from our sight if we viewed it edge-on. But there is no logical inconsistency in assuming that Flatland is really two-dimensional (i.e., has zero thickness) and that a Flatlander's two-dimensional eye has a one-dimensional retina that receives light rays emitted from the one-dimensional perimeter of the objects of vision.

E.31 we have . . . a Third unrecognized Dimension called 'height.' Apart from this epilogue, there is no indication that Flatlanders have a third dimension. Nevertheless, it seems that Abbott himself regards Flatland as a space in which the inhabitants have a slight but uniform thickness, or at least a thickness that they are unable to recognize. Writing in 1897, he described Flatland as "a world of (practically) two dimensions, in which all the inhabitants are thin Triangles, Squares, Pentagons, and other plane figures, so restricted in sight and motion that they cannot look out of, or rise or fall out of, their thin, flat universe" (Abbott 1897, 29). In the 1920s, physicists began considering theories of particle physics in which our four-dimensional (space-time) world is embedded in a higher dimensional space with the extra dimensions physically quite real but too small to be seen. Such a theory, called a superstring theory, treats elementary particles as extended one-dimensional "string-like" objects rather than as dimensionless points in space-time. In a modern "translation" of *Flatland*, the theoretical physicist Michael J. Duff gives a non-technical account of the adventures of a superstring theorist, A. Square, who inhabits a ten-dimensional world and is initially reluctant to accept the existence of an eleventh dimension. (Coincidentally, ten is the greatest dimension in the clouds on *Flatland*'s original cover.) (Duff 2001)

unrecognized Dimension, called by no name at present, but
which I will call 'extra–height'. But we can no more take cog-
35   nizance of our 'height' than you can of your 'extra–height'.
Even I – who have been in Spaceland, and have had the privilege of understanding for twenty–four hours the meaning
of 'height' – even I cannot now comprehend it, nor realise it
by the sense of sight or by any process of reason; I can but
40   apprehend it by faith.

"The reason is obvious. Dimension implies direction,
implies measurement, implies the more and the less. Now,
all our lines are *equally* and *infinitesimally* thick (or high,
whichever you like); consequently, there is nothing in them
45   to lead our minds to the conception of that Dimension. No
'delicate micrometer' – as has been suggested by one too
hasty Spaceland critic – would in the least avail us; for we
should not know *what to measure, nor in what direction*. When
we see a Line, we see something that is long and *bright*;
50   *brightness*, as well as length, is necessary to the existence of
a Line; if the brightness vanishes, the Line is extinguished.
Hence, all my Flatland friends – when I talk to them about
the unrecognized Dimension which is somehow visible in a
Line – say, 'Ah, you mean *brightness*': and when I reply, 'No, I
55   mean a real Dimension,' they at once retort 'Then measure it,
or tell us in what direction it extends': and this silences me,
for I can do neither. Only yesterday, when the Chief Circle
(in other words our High Priest) came to inspect the State
Prison and paid me his seventh annual visit, and when for
60   the seventh time he put me the question, 'Was I any better?'
I tried to prove to him that he was 'high,' as well as long
and broad, although he did not know it. But what was his
reply? 'You say I am "high"; measure my "high–ness" and
I will believe you.' What could I do? How could I meet his
65   challenge? I was crushed; and he left the room triumphant.

"Does this still seem strange to you? Then put yourself in a
similar position. Suppose a person of the Fourth Dimension,
condescending to visit you, were to say, 'Whenever you open
your eyes, you see a Plane (which is of Two Dimensions) and
70   you *infer* a Solid (which is of Three); but in reality you also see

E.33. a Fourth unrecognized Dimension. C. Howard Hinton speculates about the nature of our world, given that four-dimensional space exists. He considers two possibilities: First, we exist as three-dimensional objects in four-dimensional space, in which case he says that we are "mere abstractions" and therefore must exist only in the mind of a being that conceives us. Second, we are four-dimensional creatures, but our extension into the fourth dimension is so slight that we do not perceive it (Hinton 1886, 30). Later, he concludes even more strangely, "We must be really four-dimensional creatures, or we could not think about four dimensions" (Hinton 1888, 99).

E.35. we can no more take cognizance of our 'height.' The Square correctly notes that Flatlanders are unable to detect one another's height because they are all "equally and infinitesimally thick." (By "infinitesimally thick," he intends "extremely thin," not "infinitely thin.") Even if Flatlanders had substantial thickness, whether uniform or not, they would not be aware of it. Indeed, the very idea of "thickness" or "height" would not occur to them (Benford 1995, xv). Abbott has introduced this "unrecognized dimension" as a metaphor for the presence of spirit, or a "spiritual dimension." (But, as we have already observed in Note 19.122, he did not believe that spirits actually exist in the fourth dimension.) Abbott's use of a spatial dimension as a figure for "spiritual depth" has antecedents in the dialogues of Plato, the first person to argue that there is a "depth" to all things that is not perceived by the senses. In his commentary on the *Meno*, Jacob Klein draws an image of Meno's soul as one lacking the dimension of depth, which makes learning possible: "Meno's soul is indeed nothing but 'memory,' an isolated and autonomous memory, similar to a sheet or to a scroll covered with innumerable and intermingled characters, something of a two-dimensional and shadow-like being." Klein finds further support for this image of the soul in the *Republic* as well as the *Timaeus* (Klein 1989, 186–192).

E.40. I can but apprehend it by faith. This key statement is much more explicit than any other profession of faith in the text. Elsewhere, Abbott wrote that man cannot comprehend "absolute reality" but only apprehend it by faith (Abbott 1886, 369). In the last sentence of his letter to *The Athenaeum* (a sentence not included in this epilogue), the Square says that he apprehends by faith the truth of an unseen dimension and that he daily endeavors to inculcate this truth upon others (see Appendix A2, Footnote 9).

E.42. the more and the less. In the writings of Plato and Aristotle, "the more and the less" means continuous variation of magnitude or degree.

E.63. high-ness. "Highness" once meant the condition of being high. This condition is now called height, and "highness" has become a title of honor given to royalty. Note the irony in the High Priest's denial of the existence of his own "high-ness."

(though you do not recognize) a Fourth Dimension, which is not colour nor brightness nor anything of the kind, but a true Dimension, although I cannot point out to you its direction, nor can you possibly measure it.' What would you
75   say to such a visitor? Would not you have him locked up? Well, that is my fate: and it is as natural for us Flatlanders to lock up a Square for preaching the Third Dimension, as it is for you Spacelanders to lock up a Cube for preaching the Fourth. Alas, how strong a family likeness runs through
80   blind and persecuting humanity in all Dimensions! Points, Lines, Squares, Cubes, Extra–Cubes – we are all liable to the same errors, all alike the Slaves of our respective Dimensional prejudices, as one of your Spaceland poets has said—

'One touch of Nature makes all worlds akin.'"[1]

85   On this, point the defence of the Square seems to me to be impregnable. I wish I could say that his answer to the second (or moral) objection was equally clear and cogent. It has been objected that he is a woman–hater; and as this objection has been vehemently urged by those whom Nature's decree has
90   constituted the somewhat larger half of the Spaceland race, I should like to remove it, so far as I can honestly do so. But the Square is so unaccustomed to the use of the moral terminology of Spaceland that I should be doing him an injustice if I were literally to transcribe his defence against this charge.
95   Acting, therefore, as his interpreter and summarizer, I gather that in the course of an imprisonment of seven years he has himself modified his own personal views, both as regards Women and as regards the Isosceles or Lower Classes. Personally, he now inclines to the opinion of the Sphere that the
100   Straight Lines are in many important respects superior to the Circles. But, writing as a Historian, he has identified himself (perhaps too closely) with the views generally adopted by Flatland, and (as he has been informed) even Spaceland,

[1] The Author desires me to add, that the misconception of some of his critics on this matter has induced him to insert in his dialogue with the Sphere, certain remarks which have a bearing on the point in question, and which he had previously omitted as being tedious and unnecessary.

E.72. a Fourth Dimension, which is not colour. In his essay, "The number of dimensions of space," Hans Reichenbach substitutes color for the fourth dimension in an attempt to enable his reader to "visualize" four-dimensional space (Reichenbach 1958, 280–283).

E.73. true Dimension. The Square understands that color and brightness may be regarded as dimensions; however, by a "true Dimension" he means a spatial dimension.

E.83. all alike Slaves of our respective Dimensional prejudices. The inhabitants of spaces of every dimension all believe that the space of their experience is the only possible space. Abbott uses the slave metaphor in an earlier work: "It is the veil of our fleshly and earthly prejudice... [W]e cannot help at times being the slaves of our senses, attaching too much importance to 'the things that do appear,' too little to unseen things" (Abbott 1877a, 406).

E.84. One touch ... all worlds akin. This quotation is a slight variation on a line from Shakespeare's *Troilus and Cressida* 3.3, where Ulysses expresses the opinion that the love of novelty is common to all mankind:

> One touch of nature makes the whole world kin,
> That all with one consent praise new-born gawds,

E.88. woman-hater. We are not aware of any contemporary review that seriously accuses the Square of being a woman-hater. In his review, Robert Tucker facetiously suggests that *Flatland* "is an *ex parte* description by a Square who may at some time have suffered a disappointment at the hands of a lady" (Tucker 1884, 77).

Footnote. The additional dialogue is found in lines 16.103 through 16.143 and 19.132 through 19.148.

Historians; in whose pages (until very recent times) the des-
105 tinies of Women and of the masses of mankind have seldom
been deemed worthy of mention and never of careful consideration.

In a still more obscure passage he now desires to disavow
the Circular or aristocratic tendencies with which some crit-
110 ics have naturally credited him. While doing justice to the
intellectual power with which a few Circles have for many
generations maintained their supremacy over immense multitudes of their countrymen, he believes that the facts of
Flatland, speaking for themselves without comment on his
115 part, declare that Revolutions cannot always be suppressed
by slaughter; and that Nature, in sentencing the Circles to
infecundity, has condemned them to ultimate failure – "and
herein," he says, "I see a fulfillment of the great Law of all
worlds, that while the wisdom of Man thinks it is work-
120 ing one thing, the wisdom of Nature constrains it to work
another, and quite a different and far better thing." For the
rest, he begs his readers not to suppose that every minute
detail in the daily life of Flatland must needs correspond to
some other detail in Spaceland; and yet he hopes that, taken
125 as a whole, his work may prove suggestive, as well as amusing, to those Spacelanders of moderate and modest minds
who – speaking of that which is of the highest importance,
but lies beyond experience – decline to say on the one hand,
"This can never be," and on the other hand, "It must needs
130 be precisely thus, and we know all about it."

E.105. destinies of Women. When Abbott wrote these words, there were very few printed works on the history of women, and there were not many more in 1929 when Virginia Woolf lamented, "The history of England is the history of the male line, not of the female" (Woolf 1967, 141).

E.109. aristocratic. Although a graduate of Cambridge and an ordained Anglican Priest, Abbott was not a member of the privileged class. At the time he wrote *Flatland*, he was the headmaster of the City of London School, a middle-class day school. He alludes to his own modest social status in lines 6.141 through 6.143: "In a word, to comport oneself with perfect propriety in Polygonal society, one ought to be a Polygon oneself. Such at least is the painful teaching of my experience."

E.119. Law of all worlds. A variation of Francis Bacon's maxim, "[S]o is the wisdom of God more admirable, when nature intendeth one thing, and Providence draweth forth another" (Bacon 1965, 98).

E.122. not to suppose. In begging his readers "not to suppose that every minute detail in the daily life of Flatland must needs correspond to some other detail in Spaceland," Abbott is hinting that much of Flatland does indeed correspond to some detail in Spaceland.

The critic Darko Suvin calls Abbott "one of the true progenitors of significant modern science fiction," by which he means the literature of "cognitive estrangement." *Flatland* "estranges" its readers from the conditions of their empirical environment by recasting these conditions in an alternative universe; it is "cognitive" because this recasting of the familiar prompts engaged readers to reconsider their own worldviews (Suvin 1979, 167; 3–10).

E.128. lies beyond experience. "[T]hese words were never intended to suggest that alleged historical facts belong to the province that 'lies beyond experience.' The phrase referred to the ultimate cause of things" (Abbott 1907a, 14).

E.130. decline to say ... "This can never be ... we know all about it." The Square closes the Epilogue by expressing the hope that his book may find its way into the hands of open-minded readers.

Continued Notes

3.1. A Flatlander's "breadth" is the width of the narrowest hallway through which he or she can pass. For example, the length of a triangle is the length of its longest side, the length of a square is the length of its diagonals, and the length of a circle is its diameter. The breadth of an equilateral triangle is the length of an altitude drawn from a vertex to a side, the breadth of a square is the length of one of its sides, and the breadth of a circle is the same as its length. In Figure 3.1, all figures have the same breadth, which is the width of the strip; no two have the same length, which is indicated by the double-arrowed line segment.

Figure 3.1. Figures with the same breadth but different lengths.

3.3. A person standing waist-deep in water would appear to the surface-beings as a figure approximately the shape of an ellipse. For a person with a 38-inch waist, this "ellipse" would have about the same area as a Flatland Priest (Circle) of diameter 12 inches. Morosoff's letter is reproduced in Ouspensky (1997, 80–83).

3.29. During the second half of the nineteenth century, developments in the sciences, most notably Charles Lyell's *Principles of Geology* and Darwin's *Origin of Species*, cast doubt on theological assumptions and the absolute truth of the Bible and thereby contributed to the "Victorian crisis of faith." However, Abbott stood among the few who welcomed Darwin's theory of evolution on religious grounds; far from rejecting evolution, Abbott regarded it as a divine program for the development of the world (Abbott 1875a, "The creation of the world").

3.46. The primary purposes of marriage in classical Athens were the production of legitimate citizens to maintain the husband's estate or household and the formation of alliances to concentrate power and wealth; hence, citizen marriage was endogamous.

Marriage between cousins was a common practice among the middle class in early Victorian England. Abbott's own parents were first cousins, as were the parents of Lewis Carroll. Queen Victoria married her first cousin, Albert, the son of a German duke. Charles Darwin was the offspring as well as the contractor of a cousin marriage.

3.48. He cites the example of a giraffe, which stretches its neck to browse on the leaves of tall trees and thereby lengthens its neck; then it passes on the "lengthened-neck trait" to its offspring (Lamarck 1984, 122). The idea that acquired characteristics could be inherited was not seriously challenged by Darwin or any other biologist until late in the nineteenth century.

3.73. In several ways, Flatland serfs more closely resemble helots, the state-owned serfs of ancient Sparta, than the serfs of medieval England: They are regarded as subhuman and are brutally repressed. They far outnumber the freemen, and their rulers are preoccupied with averting a rebellion.

Although the Isosceles triangles resemble the serfs of Sparta, they represent Victorian England's substantial underclass, which William Booth (founder of the Salvation Army) described as "a vast despairing multitude in a condition nominally free but really enslaved" (Booth 1890, 23).

5.5. A Flatland observers can see at most three sides of a regular pentagon or hexagon, and at most four sides of a regular heptagon or octagon. In general, they can see at most n sides of a regular polygon with $2n$ sides, and at most $n + 1$ sides of a regular polygon with $2n + 1$ sides. The number of visible sides depends on the observer's position and distance relative to the polygon being observed. In Figure 5.1, observers in regions that are shaded alike see the same the number of sides of the hexagon.

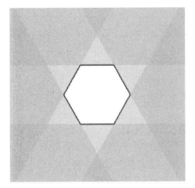

Figure 5.1. An illustration of the number of visible sides of a hexagon.

7.101. Instead of rejecting philanthropy, Abbott argued that the historical means of charity, alms-giving, ought to be replaced by a more extensive and meaningful program of assistance including "the enlargement of fields of labour, whether at home or abroad; the encouragement and inculcation of temperance; the encouragement of self-respect by providing for the poor houses and homes fit for human beings to live in; the encouragement of intellectual tastes by providing sources of legitimate enjoyment and recreation; and lastly the systematic education of the young, which, if rightly conducted, ought to stimulate at once self-respect, intelligence, temperance and industry" (Abbott 1875a, 129–130).

11.18. Hinton further explores the scientific and technological implications of planar existence in *An Episode of Flatland*, the story of beings who live not on a flat surface but on the circumference of a circular disc/planet called Astria. The universe Hinton describes is more closely analogous to our three-dimensional space than is Flatland. For example, Astrians, who have two arms and two legs, walk on the one-dimensional perimeter of their circular planet just as we walk on the two-dimensional surface of our spherical globe. Their universe has gravity (the force of gravity varies inversely with the distance instead of with the square of the distance), and Astria and a companion planet orbit a two-dimensional sun (Hinton 1907).

A. K. Dewdney made the first modern, systematic attempt to describe the general properties of a two-dimensional universe in *The Planiverse* (2001). Subtitled *Computer Contact with a Two-Dimensional World*, the book introduces Yendred, a being who makes contact with a class of computer science students who have modeled a two-dimensional world on their computer. In subsequent (computer) encounters with the students, Yendred gives a detailed description of his two-dimensional planet and its technology.

11.40. A modern mathematician would complete the argument by treating the circumference of the circle as the common limiting value of the sequences of lengths of inscribed and circumscribed polygons. But Archimedes does not use this method, probably because it involves a change of type of geometrical object – from (rectilinear) polygons to the (curvilinear) circle. Instead, he is content to prove (as it would be expressed in the language of modern mathematics) that $3\frac{10}{71} < \pi < 3\frac{1}{7}$ (Grattan-Guinness 1997, 66–67).

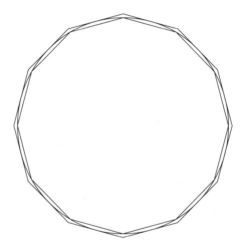

Figure 11.1. A circle with inscribed and circumscribed dodecagons.

The metaphorical use of a polygonal approximation to a circle is common in the literature. For example, Nicholas of Cusa (c. 1440) illustrates that "the precise truth is incomprehensible" with an analogy: The intellect is to truth as an approximating polygon is to the circle being approximated (Hopkins 1985, 52).

11.93. In the late-Victorian period, these schools became "institutions inculcating 'muscular Christianity,' the cult of athleticism, and patriotism, which together encouraged a distinct anti-intellectualism" (Tucker 1999, 201). John Honey gives a graphic description of the harsh means by which the schools achieved these ends in his *Tom Brown's Universe* (Honey 1977, 194–222). Commenting on the practices followed at the public schools, a writer in *The Times* (9 October 1857) suggests, "Parents may well abstain from looking too closely into the process and content themselves with the results." Several authors have noted similarities between these schools and the brutal Spartan system of education, in which boys were taken from their homes at age seven and subjected to an austere way of life.

Abbott repeatedly stressed the importance of the family in the education of children. He believed that the best education for a boy was obtained not at a public school but in a good home and a good day-school near his home. The prospectus of the City of London School declares that the school's object is to educate boys "without the necessity of removing them from the care and control of their parents" (Schools Inquiry Commission 1868b, 278).

16.197. A cube who wishes to reveal his shape to the Square could visit (pass through) Flatland as the Sphere did, manifesting himself to the Square as slices (cross-sections). As the cube enters Flatland "corner first," the initial slice is a single point, which expands to form a small triangle. One-third of the way through, the slice is an equilateral triangle containing three vertices of the cube; halfway through, the slice is a regular hexagon with its vertices at the midpoints of six of the edges of the cube. The second half of this "slicing sequence" is a reversal of the first half.

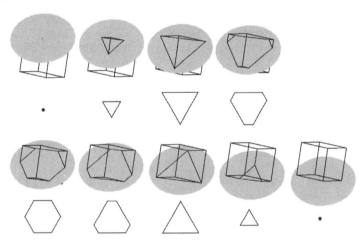

Figure 16.3. A cube passing through Flatland.

As the Square observes the slicing sequence, he is not seeing a cube but only some of its aspects. To imagine a cube, he must somehow "consolidate" this sequence of visible images to form an ideal object in his mind.

16.209. In his cave parable, Plato uses an analogy to illustrate the human condition: the cave is to the world of sense perception ("the visible realm") as the world of sense perception is to the world of thought ("the intelligible realm"). In *Flatland*, Abbott has recast Plato's analogy in the language of geometry by substituting for the cave and its prisoners a two-dimensional universe populated by geometric figures. Taken together, the Square's "visit" to Lineland and the Sphere's visit to Flatland provide an illustration of a dimensional analogy: the relationship between one-dimensional space and two-dimensional space is analogous to the relationship between two-dimensional space and three-dimensional space.

Using a dimensional analogy is an intuitive approach to the study of higher dimensions, which is often fruitful in suggesting generalizations of low-dimensional results. "If we truly understand a theorem in plane geometry, then we should be able to find one or more analogies in solid geometry, and conversely, solid geometry theorems will often suggest new relationships among plane figures. Theorems about squares should correspond to theorems about cubes or square prisms. Theorems about circles should be analogous to theorems about spheres or cylinders or cones. But if we learn a good deal by going from two dimensions to three, would we not learn even more by going from three dimensions to four?" (Banchoff 1990a, 8)

16.211. Aristotle denies the possibility of extending the derivation sequence beyond the third dimension: "We cannot pass beyond body to a further kind, as we passed from length to surface, and from surface to body. For if we could, it would cease to be true that body is complete magnitude. We could pass beyond it only in virtue of a defect in it; and that which is complete cannot be defective, since it has being in every respect" (*On the Heavens*, 268ab).

References to point, line, plane surface, and solid as the constituent elements in ideal and physical magnitudes first occur in the fourth century BC. Aristotle's account in the *Metaphysics* (985b–988a) suggests that the problem of the derivation of magnitudes was a live contemporary problem. "Plato, Speusippus, and Xenocrates, each in differing ways, derive magnitudes from the sequence point, line, surface, solid, and their speculations give rise to a lively controversy concerning indivisible lines" (Philip 1966, 32).

19.130. He is completely unaware that his revelation is incomprehensible to any Flatlander. Mere words could never convince the Square that three-dimensional space exists, but the Sphere's argument that Flatlanders could not see one another unless they have a third spatial dimension (line 16.123) and his argument from analogy (lines 16.207 through 16.286) are not just futile, they are fallacious. The Sphere has so little empathy for the Square that he does not recognize that his own rejection of even the possibility of a fourth dimension mirrors the Square's adamant refusal to consider the possible existence of a third dimension.

19.224. Oresme would not "construct" the hypertetrahedron, or 5-cell, by allowing each point of the tetrahedron to move in the direction of a fourth dimension because "it does not happen that a fourth dimension (*quartam dimensionem*) exists or is imagined." Instead, he would partition the tetrahedron into an infinite family of triangular slices and then "construct" a tetrahedron on each slice (Oresme and Clagett 1968, 175–177, 531).

Appendix A: Critical Reaction to Flatland

Flatland was reviewed in some of the best "papers," and the reviews were largely favorable. Two exceptions were brief notices in *The Times* ("How any man should have apparently spent labour and ingenuity on an elaborate scientific joke that must fall as flat as anything in his own 'Flatland' is one of the mysteries that no one may fathom") and the *New York Times* ("Some little sense is apparent in an appeal for a better education for women, but beyond that all the rest of Flatland is incomprehensible"). The most prescient reviews were the ones in *The Literary World* (London) ("This cleverly elaborated fancy is not only likely to create a present sensation in the thinking world, but also to find an abiding place in the classic domains of the great satires of history") and *The Literary World* (Boston) ("With so much wit and grace of style is this clever satire on the limits of our knowledge wrought out that we shall not be surprised to see it take a permanent place in literature").

A1. Contemporary reviews

The Oxford Magazine (5 November 1884, 387)
The Academy (8 November 1884, 302)
The Literary World (London) (14 November 1884, 389–390)
The Architect (15 November 1884, 326–327)
The Athenaeum No. 2977 (15 November 1884, 622)
Nature (27 November 1884, 76–77)
Knowledge 6 (28 November 1884, 449)
The Spectator (29 November 1884, 1583–1584)
The University Pulpit: Supplement to The Cambridge Review 6 (1884/1885, lxxi–lxxii)
The Times (25 December 1884, 6)
Light 5 (10 January 1885, 16–17)
American 9:237 (21 February 1885, 312)
New York Times (23 February 1885, 3)

New York Tribune (6 March 1885, 6)
The Literary World (Boston) (21 March 1885, 93)
The Nation 40 (26 March 1885, 266–267)
The Literary News (March 1885, 85; reprinted from the *Boston Advertiser*)
Science (27 February 1885, 184; 3 April 1885, 265–266)
The Critic (18 April 1885, 18)
The Overland Monthly (April 1885, 446)
Lippincott's Magazine (May 1885, 528)
The Unitarian Review 24 (July 1885, 90)
City of London School Magazine 8 (December 1885, 217–221)

A2. *The review of* Flatland *in* The Athenaeum

The most important review of *Flatland* appeared in *The Athenaeum*, which was at the time the most authoritative scientific and literary weekly. As we demonstrate below, this review led directly to the "second and revised" edition of *Flatland*.

The Athenaeum No. 2977 (15 November 1884, 622)[1]

That whimsical book *Flatland* by a Square (Seeley & Co.) seems to have a purpose, but what that may be it is hard to discover. At first it read as if it were intended to teach young people the elementary principles of geometry. Next it seemed to have been written in support of the more transcendental branches of the same science. Lastly we fancied we could see indications that it was meant to enforce spiritualistic doctrines, with perhaps an admixture of covert satire on various social and political theories. The general purport of it is to show how a being shaped like a square, born and bred in a world in which everything took place on a plane surface, and where consequently only two dimensions were conceived, obtained by a sort of revelation knowledge of a third dimension. He has previously in a dream studied the conditions of existence in a world of one dimension, where everything is a line or point, and nobody can pass any one else. There is some ingenuity in the way in which these conceptions are worked out, but it is rather spoilt to the mathematical mind by the conception (which, indeed, was unavoidable) of lines and points as objects which can be seen. **Of course, if our friend the Square and his polygonal relations**

[1] *The Athenaeum* Collection at City University (London) includes issues that were "marked" by the editor of the day with the reviewers' names. From this source, we learned that the anonymous reviewer of *Flatland* was Arthur John Butler, an accomplished Italian scholar widely known for his translations and contributions to the study of Dante. Butler graduated from Trinity College, Cambridge, B.A. (eighth classic and nineteenth junior optime in mathematics) 1867 and M.S. 1870. He was a regular contributor to *The Athenaeum* for thirty-five years.

could see each other edgewise, they must have had *some* **thickness, and need not, therefore, have been so distressed at the doctrine of a third dimension.**[2] There is something rather funny in the idea that a being of *n* dimensions, when addressed by a being of *n* + 1, fancies the voice which he hears to proceed from his own inside; but no doubt it is in strict harmony with facts, and probably represents what we should all of us feel if we got into a region where it was possible to tie a knot in a closed loop of string, as it is in the world of four dimensions. When we saw the feat performed we should doubtless be as much surprised as our Square was when the Sphere told him the contents of his house without opening the door or taking off the roof. If we came back and told about it, we should, equally without doubt, fare much as the unlucky narrator of this history did.

The Athenaeum No. 2978 (22 November 1884, 660)

Literary Gossip. That curious little book 'Flatland,' which we noticed last week is said to be the production of the head master of a well-known school.[3]

The Athenaeum No. 2980 (6 December 1884, 733)

THE METAPHYSICS OF FLATLAND
Flatland, the State Prison, Nov. 28, 1884.

I write from a world that has been truly and literally described as "weary, stale, *flat*, and unprofitable," – from the land of Two Dimensions, some of the characteristics of which I have recently endeavoured to describe in a little treatise entitled 'Flatland.'

Into the dimness of my dull existence in this region there has penetrated a notice of my work which appeared in a recent number of the *Athenaeum*, and which raises a neat question – shall I say metaphysical or psychological? – which may possibly interest your readers.

Your not unfriendly, but, as I venture to think, too hasty critic, while complimenting me on the "ingenuity" of my simple description of my native land, and while admitting that the incidents recorded in my history, though "funny," are nevertheless "strictly according to facts," has, nevertheless, cast an implied censure on my intelligence, and on that of my

---

[2] We have set in boldface type the portions of the review and the Square's letter that have direct connections to the text of *Flatland*. The reviewer's assertion that a Flatlander ought to realize that he and his fellow countrymen have a third spatial dimension (thickness), else they would not be visible to one another, is the origin of the important differences between the first and second editions. Abbott devotes two-thirds of the Epilogue to refuting the assertion, yet he puts it in the mouth of the Sphere at line 16.103 and in the mouth of Square at line 19.132.

[3] *The Athenaeum*'s editor attributes this item to A. J. Butler.

countrymen, by declaring that, though we think we are of Two Dimensions, we are really of Three, and ought to know it. The narrative is spoilt, he says, "for mathematical minds," because any *visible* line must really have thickness as well as length; and therefore all our so-called plane figures, besides having length and breadth, must really have some degree of thickness, or height – in other words a Third Dimension; and of this, he implies, we ought not to be ignorant.

I admit your critic's facts, but I deny his conclusions. It is true, no doubt, that *we* really have a Third Dimension, just as it is also true that *you* have a Fourth. But just as you are not aware that you belong to the Fourth Dimension, so neither are we aware, nor can we be made logically aware, that we belong to the Third.[4]

A moment's reflection will make this obvious. **Dimension implies measurement. Now, our lines are so thin that they cannot be measured. Measurement implies degrees, the more and the less; but all our lines are equally and infinitesimally thin, or thick, whichever you please to call it; so that we in Flatland can neither measure their thinness, nor even take cognizance of it. Where you speak of a line as being long and thick (or thin), we speak of it as being long and *bright*;**[5] "thickness" (or "thinness") never enters our heads, and we do not know what you mean by it. I knew what it meant once, during the few hours I spent in Spaceland; but I cannot realize it now. I take it on trust; but I cannot now make a mental image of it even to myself, much less to my countrymen.

Does this puzzle you? Then put yourself in my place. Suppose a being of the Fourth Dimension, condescending to visit you, were to address you thus: "You creatures of Three Dimensions *see* a plane (which is of Two Dimensions) and you *infer* a solid (which is of Three); but in reality what you call a plane has another Dimension of a kind not known to you"; what would you reply? Would you not call for a policeman to see your visitor safely locked up in some asylum?[6]

Well, precisely this has been my reception when I have attempted to demonstrate the facts insisted on by your critic. **Only yesterday, when the Chief Circle (in other words the Chief Priest) paid his annual visit to my prison, I endeavoured to prove to him that the Figures which we saw around us had a Third non-recognized Dimension, being not only long and broad, but also what you in Spaceland call "high." What was his reply? Simply this: "Dimension implies measurement. You say I am 'high'; measure my 'high-ness' and I will believe you." I was crushed, and he left the room in triumph.**[7]

---

[4] Compare to the Epilogue, lines 30 through 35.
[5] Compare to the Epilogue, lines 41 through 49.
[6] Compare to the Epilogue, lines 66 through 75.
[7] Compare to the Epilogue, lines 57 through 65.

Sir, I am a humble Square, and I do not deny the superiority of your critic, who is doubtless a Cube; I impugn neither the exactness of his mathematics nor the regularity of his proportions; in the language of Spaceland, I am ready to admit that he is "a regular Cube and no mistake." But I respectfully submit that his knowledge of human nature is not equal to his knowledge of mathematics. He has forgotten that **we are all alike – Points, Lines, Squares, Cubes, Extra-Cubes, whether of no Dimensions or of many Dimensions – liable to the prejudices of our several Dimensions, brothers in error; as one of your own poets also has said, "One touch of nature makes the world akin,"**[8] meaning thereby not one world only, but all worlds, and not excepting the favoured world of Three Dimensions. And I must say I take it ill that I should be, however gently, censured for appearing to be ignorant of a truth which I firmly apprehend by faith, and which I daily endeavour to inculcate upon others.[9] A Square.

*   *

*

If we understand the Square rightly, all that is wanting to make the Flatlanders realize a third dimension, and to settle circularism once for all, is a delicate micrometer. For he seems to admit that the edges of himself and his countrymen really are extended surfaces – as, indeed, appears from the fact which he elsewhere mentions, that they were capable of receiving colour. He is not, therefore, in the same position with regard to the third dimension as we of this world with regard to a fourth. The truth is, it may be suspected that our Square, having once in some measure grasped the conception of three-dimensioned space, cannot now wholly divest himself of it. He thinks, so to speak, in three dimensions. For instance, he talks in one place of hearing **the sound of his wife's retreating footsteps,**[10] a bold metaphor indeed to apply the motion of a line on a plane. But, with a degree of intellectual insincerity probably unconscious, certainly pardonable in a person situated as he is, he thinks it necessary to persist in saying that he apprehends by faith a truth which he has really learnt from the evidence of eyesight; thus making a serious confusion between the functions of faith and sense. The Square does his reviewer too much honour in supposing him to be a regular cube. The best he can claim to be is a rectangular parallelepiped; and he finds it hard enough to live up to that configuration in space of the kind he knows, so that he is content to do without speculations as to the ways of beings in worlds of more or fewer dimensions.

[8] Compare to the Epilogue, lines 79 through 84.

[9] In the Preface to *Apologia*, Abbott says that the message of 2 Corinthians 4:18 ("The things which are seen are temporal; but the things which are not seen are eternal") is the basis for all that he tries to impart to others (Abbott 1907, xii).

[10] The Second Edition corrects this mistake at line 16.1.

A3. Another letter from the Square

In addition to his letter to *The Athenaeum*, the Square sent a letter to *Knowledge*, a journal edited by Richard A. Proctor, who was a contemporary of Abbott at St. John's College, Cambridge. This second letter may have been a belated response to an essay in which Proctor declared that the idea of a fourth dimension of space was absurd as well as inconceivable, and that it was "idle to talk of creatures having only length and breadth and what such unimaginable creatures would think of our comfortable triple dimensions of length, breadth, and thickness" (Proctor 1884, 44). Note that Proctor's acknowledgement of the Square's letter (which he did not publish), given here, closes by gently mocking the Square's propensity for alliteration.

> "'A Square' sends a 'wail from Flatland,' or in other words, the well-known Tridimensional, whose romances of Flatland have touched many students of quadri-dimensional mathematics, has sent a letter describing the anguish of 'A Square' imprisoned in such cells as Mr. Garbett failed to find escape from. We would publish this piteous lamentation were more of our readers quadri-dimensional; but the number of such readers is so limited, that we must ask the ingenious author to forgive us if we refrain. Will he kindly regard his letter printed in planes persistently perpendicular to those of the pages of *Knowledge*" (Proctor 1887).

A4. James J. Sylvester and **Flatland**

Of the thousands of mathematics teachers who have recommended *Flatland* as an introduction to higher-dimensional geometry, we can be quite sure that the first was the great English mathematician, James J. Sylvester. In late 1883, Sylvester left Johns Hopkins University and returned to England to become the Savilian Professor of Geometry at the University of Oxford. At New College, Oxford, in the fall of 1884 he was lecturing three days a week on analytic geometry to a "numerous and interesting class." On November 2nd, he wrote to his friend Arthur Cayley that he had recommended to his class that they "procure Flatland (by Abbott of the City of London School) in order to obtain a general notion of the doctrine of space of *n* dimensions" (Sylvester 1884). Sylvester's letter to Cayley shows that he had (or at least had access to) a very early copy of the book. The first review of *Flatland* appeared in *The Oxford Magazine* (5 November 1884), and we conjecture that Sylvester was its author. In response to our inquiry, Sylvester's most recent biographer, Karen Parshall, said that there is "a good possibility" that Sylvester wrote the *Oxford* review. She notes that it is peppered with French phrases, which Sylvester favored.

Appendix B: The Life and Work of Edwin Abbott Abbott

Appendix B1. Abbott chronology

There is no full-scale biography of the author of *Flatland*, Edwin Abbott Abbott. The entry for Abbott in the *Dictionary of National Biography*, written by his student Lewis Farnell and revised by Rosemary Jann, supplies a valuable outline of his life. A. E. Douglas-Smith, who had access to Abbott's weekly letters to Howard Candler as well the autobiographical, "Confessions of a rationalist Christian" (both now lost), gives a good account of Abbott as headmaster in Douglas-Smith (1965). Former students have given compelling testimony to his greatness as a teacher (see Conway (1926); Farnell (1926); Steggall (1926)). Abbott expounds his liberal theology in several popular works, for example, *Philochristus*, *The Kernel and the Husk*, and *Through Nature to Christ*. William Sanday provides an overview of Abbott's prodigious biblical scholarship in his reviews of Abbott's books in *The Times Literary Supplement*. A biographical sketch of even modest length is beyond our scope; instead we offer this chronology, which provides a framework of Abbott's life and work.

1838. Edwin Abbott Abbott is born in London on December 20. He was the fifth child and oldest son of Edwin Abbott (1808–1882) and Jane Abbott Abbott (1806–1882), who were first cousins. His father was "a most efficient and popular" headmaster of the Philological School, Marylebone, for forty-five years. He was one of the first to advocate a more thorough English training in the schools. He was the author of *A Second Latin Book* (1858) ("which for its clear method and copious collection of idioms is even now valuable"), *Greek Tragic Iambics* (1864), and *A Concordance to the Works of Alexander Pope* (1875), as well as "several admirable school-books, written and cheaply printed for his own pupils at a time when good and cheap school-books were hardly to be obtained" (*Academy*, 10 June 1882, 415; *The Athenaeum*, 3 June 1882, 700; Borg 2004).

1850–1857. Abbott attends the City of London School, where he forms enduring friendships with Howard Candler, William S. Aldis, John Y. Paterson, Alfred R. Vardy, and John R. and Richmond Seeley. He was Captain of the school, 1854–1857.

Howard Candler, Abbott's closest friend.

1856. J. Llewelyn Davies becomes Rector of Christ Church, Marylebone, which the Abbott family attended. Abbott testifies to the impact that Davies' first sermon had on his life in (Abbott 1886, 9), and in 1906 he wrote to Davies, "There is no one to whom I am more indebted than to you for what I count the most precious of all possessions" (Abbott 1906).

1857. Abbott enters St. John's College, Cambridge, as a scholarship student.

1859. Abbott's first published paper, a review of Richard G. White's *The Works of William Shakespeare*, appears (Abbott 1859).

1859. In May, Edward, the older of Abbott's two brothers, died at sea along with 106 other members of the crew when the ship on which he was an assistant master was caught in a tornado and sank off the coast of Africa. Two months later, his sister Jane died following a long illness. In *The Kernel and the Husk*, Abbott recounts that a college friend gave him a copy of Tennyson's *In Memoriam* at about this time. "I read the poem again and again, and committed much of it to memory; and it exerted an 'epoch-making' influence on my life" (Abbott 1886, 10).

1860. Abbott wins the Camden Medal awarded by the University of Cambridge for the best poem in Latin hexameter verse (Camden 1860).

1861. Abbott is seventh among those gaining second class honours on the Mathematical Tripos (see Appendix B2). In the Classical Tripos,[1] taken about one month after the Mathematical Tripos, Abbott is the senior classic. He also wins the First Chancellor's Medal for classics. His St. John's College tutor in the classics was Francis France, who was the senior classic in 1840; however, like all the best men, Abbott read privately with Richard Shilleto, one of the greatest Greek scholars that England ever produced and for thirty years the leading classical coach at Cambridge. Immediately following Abbott on the Tripos list was George O. Trevelyan, who subsequently was a Member of Parliament and a minister in all of Gladstone's five governments.

1861. Abbott is editor of the St. John's *Eagle*, December 1861 to June 1862.

1862. Abbott gains first class honours in the Theological Tripos. On the basis of their papers on the Theological Tripos, Abbott and R. C. W. Raban are judged equal for the Scholefield Prize awarded to the student who demonstrates the most complete knowledge of the Greek Testament and of the Septuagint Version of the Old Testament.

1862. Abbott is elected a fellow of St John's. His friend William S. Aldis, the senior wrangler, would ordinarily have been assured of a fellowship as well, but becoming a fellow entailed subscribing to the thirty-nine Articles of the Church of England, and so Aldis, a Baptist, was ineligible. Instead, he spent the next ten years as a private mathematical tutor at Cambridge. In 1871, he became Professor of Mathematics and later Principal of the College of Physical Science at Newcastle. He left England in 1884 to become Chairman of the Department of Mathematics at Auckland College in New Zealand, where he remained until his retirement in 1896. His books, *Solid Geometry* (1865) and *Geometrical Optics* (1872), long remained standard works (*The Times*, 13 March 1928, 21).

  Years later, in summing up a student debate of the question whether all the privileges of the universities should be opened to Nonconformists, Abbott remarked, "Of course as a clergyman of the Church of England I cannot agree with all that Asquith[2] has said. But one thing I cannot forget: My friend Aldis and I went up together from this school to

---

[1] The Classical Tripos consisted of eleven three-hour sessions spread over six days. Six sessions were devoted to translations from Latin or Greek prose or verse into English. Four sessions dealt with translations of English prose and verse into Latin or Greek, and the final session tested students' knowledge of the history of Greece and Rome.

[2] Herbert H. Asquith was one of Abbott's favorite students (see 1909 below). Abbott always presided over the monthly student debates at CLS but used the time to correct or mark exercises, except when Asquith was speaking. Then he devoted his whole attention to the debate, not for the sake of the subject but for Asquith's eloquent rhetoric (Garnett 1932).

Cambridge. He was senior wrangler and I senior classic in the same year. As a member of the Church of England, everything was open to me. As a Nonconformist, everything was closed to him" (Aldis 1940). It was not until 1871 that the University Tests Act opened all lay posts at the universities to men of all creeds.

Abbott alluded to their time together at the City of London School as well as their triumph on the Tripos examinations (taken in the Senate House) when he inscribed a first edition of *Flatland* to his friend:[3]

W. S. Aldis

from the Author

in memory of the days when we worked

together at School & in the Senate House

Oct. 1884.

1862–1864.   Abbott is composition master at King Edward's School, Birmingham, where he forms a lasting friendship with J. Hunter Smith.

1862.   Abbott is ordained deacon by the Bishop of Ely on June 15th.

1863.   Although his health was never good, Abbott liked playing tennis; he sculled, swam, and was a great walker (Abbott 1906; Douglas-Smith 1965, 237). On a walking tour of Derbyshire, he and a friend stayed the night with Henry Rangeley of Unstone Grange, and he soon fell in love with his host's eldest daughter, Mary Elizabeth. At the time of their marriage in July 1863, Abbott was a fellow at St. John's. Celibacy was a condition for the tenure of a fellowship at Cambridge until 1882, and so getting married entailed resigning his fellowship.

1863.   Abbott is ordained priest by the Bishop of Worcester on December 20th.

1864.   Abbott is awarded the M.A. degree.

1864.   Abbott reads David F. Strauss's *The Life of Jesus*, in which miraculous elements in the gospels are characterized as "mythical." Concerning this controversial book, Abbott wrote, "If I had read Strauss' *Life of Jesus* three years before, instead of three years after, taking my degree, I should have, in all probability, at once cast off faith in Christ" (Abbott 1877a, 21).

1864–1865.   Abbott is an assistant master at Clifton College, Bristol, under the distinguished headmaster, John Percival, who was later bishop of

---

[3] Aldis's copy of *Flatland* is now part of the Hay Star rare book collection of the Brown University Library, a gift of Thomas F. Banchoff.

Mary Elizabeth Rangeley Abbott.

Hereford. Percival once said, "Had Edwin Abbott been able to continue preaching, he would have been the greatest preacher in the English Church" (Obituary 1926a).

1865. Abbott becomes headmaster of the City of London School. His appointment was extraordinary because he was only twenty-six years old; most of his assistant masters had taught him when he was a boy, and at least one had been passed over in his favor for the post. He looked so young that he was sometimes mistaken for a member of his sixth form, but he soon established his authority over the boys and won the loyalty and admiration of the staff. He later wrote, "Though I was somewhat young for the work, I had sense enough to make no changes all at once, nor to frighten people by talking about changes" (Douglas-Smith 1965, 159–160).

1866. John R. Seeley's *Ecce Homo: A Survey of the Life and Work of Jesus Christ* appears. Abbott dedicated *Philochristus* to "the Author of 'Ecce Homo,' not more in admiration of his writings than in gratitude for the suggestive influence of a long and intimate friendship" (Abbott 1878).

1868. Abbott's only son, Edwin (d. 1952), is born. He attended St. John's Wood School, St. Paul's School, and Caius College, Cambridge (1886–1890). In 1889, he took a first class in Part I of the Classical Tripos; the following year he took only a second class in Part II, but nonetheless won the Chancellor's Medal. In 1890, he was elected to a Fellowship at Jesus College and appointed Lecturer in Classics. In 1912, he became Senior Tutor, and he held that office until 1932, when a nervous breakdown caused him to resign his tutorship and go out of residence. In the following year, he settled in North London with his sister. "His shyness, with undergraduates, as even with older people, was a life-long characteristic; but he was devoted to his duty, and greatly beloved by all his pupils and by his colleagues... The key to the affection in which he was held throughout his long life of eighty-four years was his complete unselfishness, which made a deep impression on all who had the privilege of knowing him" (Obituary 1952).

Edwin Abbott.

1868. In his essay, "The Church and the congregation," Abbott describes the condition of the Church of England with a parable of a house in disrepair (Abbott 1868a).

1868. At a special evening service on May 3rd in the nave of Westminster Abbey, Abbott delivered a sermon, "The signs of the Church." A dignitary of the Church complained to the Lord Mayor that the whole tone of the sermon was "to set the poor against the rich," and the Mayor raised the matter at a meeting of the Court of Common Council, the administrative body of the City of London (*The Times*, 24 July 1868, 10; *The Times*, 25 July 1868, 12). Several years later, Abbott commented on this controversy: "If indeed a preacher may be fairly accused of 'setting class against class' because he cannot help betraying a feeling of profound dissatisfaction with the present state of religious feeling and religious action, then I must plead guilty to the charge. For, optimists though we may be in our hopes as to Christ's future Kingdom on earth, yet surely we are allowed, nay, bound, to feel profoundly dissatisfied with our present miserable realization of it" (Abbott 1875a, viii–ix).

1869. *Shakespearian Grammar: An Attempt to Illustrate Some of the Differences between Elizabethan and Modern English: For Use in the Schools* is published. Abbott was one of the first to apply the principles of classical scholarship to the criticism of Elizabethan English. His stated objective was "to furnish students of Shakespeare and Bacon with a short, systematic account of some points of difference between Elizabethan syntax and our own. The *words* of these authors present but little difficulty. . . But the *differences of idiom* are more perplexing" (Abbott 1870, 1). The first edition was published in July 1869; a slightly enlarged second edition followed in early 1870. The success of these two editions encouraged him to enlarge his manuscript to a complete book of reference for Shakespearean syntax and prosody. For this purpose, he reread the whole of Shakespeare and produced a third edition within a year of the publication of the first. His text remains invaluable for advanced students of Elizabethan drama.

1869. Abbott is one of the founding members of the Headmasters' Conference, and for some years its Secretary.

1870. *Bible Lessons* is published. These lessons, which Abbott gave to his fifth and sixth forms at the City of London School, are presented as dialogues between a teacher and a pupil.

1870. Women were allowed to vote in the elections and serve on the school boards that were established by the Education Act of 1870. In the first School Board election for London, Abbott worked on behalf of two women who were longtime allies in the cause of women's rights: Emily Davies, suffragist and one of the founders of Girton College, the first women's college at the University of Cambridge, and Elizabeth Garrett, the first woman to qualify as a doctor in Great Britain. Both women were elected; Garrett recorded the greatest number of votes of all candidates in the entire metropolis (The Education Act, *The Times*, 2 December 1870, 4).

1870. Abbott's daughter, Mary (d. 1952), is born. Mary was taught by her parents until 1881, when she entered South Hampstead High School, where she studied for five years. In addition, she was tutored privately by Eugénie Sellers Strong, who later became a well-known archaeologist and art historian. Mary Abbott attended Girton College (1889–1892) as a Classical Foundation Scholar. Like her father and brother, she was placed in the first class of the Classical Tripos; however, she did not receive a degree from the University of Cambridge. (At Cambridge, women were first allowed to take the Tripos examinations in 1881. They were admitted to "titular" degrees in 1921, but they were not admitted to full membership in the University until 1947.) Like one-seventh of all Victorian women, she never married (*Girton* 1948, 54).

Mary Abbott.

Mary collaborated with her father in his literary and theological work from 1892 until 1917. The reviewer of Abbott's last book refers to "the beautiful association of father and daughter in a work of exacting scholarship" (Sanday 1917). Abbott twice acknowledged the extent of his debt to his daughter. On the dedication page of his *Johannine Vocabulary*, he says, "To my daughter by whom the main materials for the work were collected and classified and the results corrected and revised this book is dedicated." In the preface to his last book, he says, "To my daughter I owe not only the Indices at the end of the volumes, but also a close and searching recension of the whole work, which has detected innumerable faults, and has gone far to remedy the author's increasing infirmities of memory and defects in exact and accurate expression" (Abbott 1917, xx).

1871. From the beginning of his tenure at the City of London School, Abbott reformed the traditional curriculum to include the study of English; all students were required to study of the works of Milton, Shakespeare, and Sir Walter Scott (*The Times*, 31 July 1869, 5). In essays and lectures, Abbott made a case for the teaching of English and gave detailed examples of his own methods (Abbott 1868b; 1871). He emphasized that the study of English ought not to be a mere study of words but a study of thought. Abbott and Seeley's *English Lessons for English People* marks the beginning of what has been described as "the first modern phase in the teaching of English" (Michael 1987, 2).

A number of students in Abbott's sixth form became famous in the world of English letters: A. H. Bullen, literary editor and publisher; H. C. Beeching, author and Dean of Norwich; Sir Sidney Lee, literary scholar and editor of the *Dictionary of National Biography*; C. E. Montague, journalist and novelist; Sir Israel Gollancz, literary scholar and a cofounder of the British Academy; G. W. Steevens, war correspondent and writer. His greatness as a teacher and headmaster was recognized not only for producing students like these, but also for his innovations in curriculum and instruction on behalf of all students. A reviewer of Abbott's *Shakespearian Grammar* said:

> Among head masters of English public schools, Mr. Abbott is honourably known for his efforts to teach his boys to think, and not to cram, and for his determination not to sacrifice those of his boys who are not meant for the University to those who are meant for it. While sending up to Oxford and Cambridge a succession of young fellows who have won brilliant honours there, Mr. Abbott has kept up in his school the most thorough teaching of English known in any of our public schools (Review 1870, 1551).

1872. *Good Voices, or a Child's Guide to the Bible* is published.

1872. Throughout the Victorian period, few schools taught English as a separate subject; instead it was expected that students would learn their native language as a by-product of the studies of Latin and Greek. Abbott believed strongly in the value of classical studies, but he was dismayed that teachers generally did not require students to express their translations in good English. To remedy this deplorable condition of English usage, he wrote a small handbook of composition, *How to Write Clearly*. The book was essentially a set of fifty-six commonly violated grammatical principles ("rules") together with illustrative examples and exercises. It was very popular on both sides of the Atlantic; the American edition published by Roberts Brothers was used extensively at Harvard University and the University of Michigan.

1872. Abbott is made Doctor of Divinity by the Archbishop of Canterbury.

1872. At the Headmasters' Conference, Abbott speaks frankly of his mistakes as a beginning teacher: "Personally, I feel that by some kind of professional training I should have been saved from many mistakes that I deeply regret, for I gained much valuable experience of teaching at the expense of my pupils. And I think many other teachers entertain with myself a feeling of regret, approaching to something more keen, at the mischief they have done their pupils from inefficiency" (Fitch 1876, 104).

1873. *Latin Prose through English Idiom* and *Parables for Children* are published.

1874. Abbott is a vice-president of the New Shakspere Society founded by Frederick J. Furnivall. At the opening meeting in March, Abbott reads a paper by Frederick G. Fleay, "On metrical tests as applied to dramatic poetry." He spends a good deal of time during 1874 in a futile attempt to mediate a protracted quarrel between Furnivall and Fleay.

1875. In January, Abbott reads his paper, "On the first two quartos of Hamlet, 1603, 1604," to the New Shakspere Society.

1875. *Cambridge Sermons Preached before the University* is published. Abbott's student J. E. A. Steggall recalls that Abbott's first University sermon was so controversial that at the second one, many of the "important" stalls at Great St. Mary's Church, Cambridge, were empty in protest (Steggall 1926, 123). This volume includes the sermon, "The signs of the Church," delivered in Westminster Abbey in May 1868.

1875. At a meeting of the National Association for the Promotion of Social Science, Abbott reads a paper, "Middle class education," in which he asserts that middle-class schools are in desperate need of improvement and urges that teachers in these schools be required to demonstrate competence in the subject matter that they teach and trained for the work of teaching (Miscellaneous 1876, 463).

1875. Abbott supplies critical remarks on Pope's style, English, and diction in a twelve-page introduction to his father's book, *A Concordance to the Works of Alexander Pope*.

1875. *How to Tell the Parts of Speech* and *How to Parse* are published.

1875. Abbott thanks George Eliot for her "kind note," which he says encouraged him to return to writing *Philochristus* after having set it aside for several years. In closing, he tells her that he has "derived more spiritual strength and more intellectual delight" from her works than from those of any other author but Shakespeare (Abbott 1875b).

1875. Abbott begins a long campaign for the removal of the City of London School to more suitable premises.

1876. *Bacon's Essays* is published, a heavily annotated edition of Bacon's essays together with an introduction to his life and thought. Abbott's unfavorable assessment of Bacon's character in this book as well as *Bacon and Essex* and *Francis Bacon* provoked a sharp rebuttal from the preeminent authority on Bacon, James Spedding. In her re-evaluation of Francis Bacon, Nieves Mathews argues strongly that the negative portrayal of Bacon by authors like Abbott and Thomas Macaulay, which has been echoed by subsequent biographers, is inaccurate (Mathews 1996).

1876. Abbott is the Hulsean lecturer at Cambridge.

1877. In a letter to the *Spectator*, Abbott maintains that there is an urgent need to train teachers for schools above the elementary level. He says that the great bulk of middle-class private schools are teaching children "to study many things and know nothing" (Abbott 1877d).

1877. *Bacon and Essex: A Sketch of Bacon's Earlier Life* is published.

1877. *Through Nature to Christ, or the Ascent of Worship through Illusion to Truth* is published. The author of Abbott's obituary in *The Times* comments that the publication of this book "brought down a storm of hostile criticism because it defined for the first time that liberal attitude to theology which marked all his subsequent work" (Obituary 1926a) (see Notes 22.23 and 22.111).

1877. Abbott is a Select Preacher at Oxford.

1878. Abbott is one of the signers of the clerical address in favor of neutrality in the Russo-Turkish war (*The Times*, 17 January 1878, 6).

1878. Abbott joins the Thirlmere Defence Association, formed to resist the attempt by the Manchester Corporation to create a large reservoir in this Lake District valley. The Association delayed but did not stop this scheme, which was completed in 1894 (*The Times*, 18 January 1878, 10).

1878. After many years of study and thought, Abbott finally publishes *Philochristus*, a retelling of New Testament narratives by a fictional disciple of Jesus, who gives voice to Abbott's own peculiar theological position.

1878. Abbott speaks at the annual general meeting of the National Society for Women's Suffrage, July 1st (*The Times*, 2 July 1878, 10).

1879. In July, Abbott attends the annual meeting of the Marylebone United Liberal Association (*The Times*, 24 July 1879, 7).

1879. *Oxford Sermons Preached before the University* is published.

1879. "Gospels" (*Encyclopaedia Britannica*, 9th ed.) is published. This book-length essay was one of the earliest presentations in English that upheld the theory that Mark was the oldest of the synoptic gospels and that it most closely approximates their presumed common source.

   The page proofs of this article, which had been heavily annotated over the years, were among Abbott's papers when he died. Although he saved almost none of the letters he received, he kept a letter from Benjamin Jowett (the master of Balliol College, Oxford), who warned him that he would be "very much attacked" but said that the article "will make an era in English theology" (Jowett 1879).

1880. W. G. Rushbrooke's *Synopticon: An Exposition of the Common Matter of the Synoptic Gospels* appears. Printed in a large folio volume, it presents the Greek texts of the Synoptic Gospels (Matthew, Mark, and Luke) in parallel columns; the words common to all three, the words common to each pair, and the words peculiar to each are distinguished by differences in type and color. The work is dedicated to Abbott, who suggested that Rushbrooke undertake it and assisted him in writing it.

Of all his students, Abbott was closest to Rushbrooke, who was a student at the City of London School (1862–1868), Captain of the school (1867–1868), and like Abbott earned first class honours in the classics at St. John's College, Cambridge. The author of Rushbrooke's obituary observed, "Abbott was a great maker of men, and in Rushbrooke he made a great schoolmaster." He served as master of the fourth form at CLS (1872–1893) and was a distinguished headmaster at St. Olave's, Southwark (1893–1922) (*The Times*, 1 February 1926, 17).

1880. Abbott is the administrator of the estate of his sister Elizabeth Mead Parry and her husband John Humffreys Parry, who die suddenly. Their son, Sir Edward Abbott Parry, remembers his uncle Edwin fondly in Parry (1932).

1880. *Via Latina: A First Latin Book* is published.

1882. *Onesimus: Memoirs of a Disciple of St Paul* is published.

1882. Abbott writes to W. E. Gladstone, imploring the Prime Minister not to hold the Egyptian people responsible for the debts incurred by the previous despotic ruler; Gladstone's response is noncommittal (Abbott 1882b; Gladstone 1882).

1883. *Hints on Home Teaching* is published, a book written to assist governesses and private tutors and especially to enable parents to contribute to the training of their children.

1883. The City of London School finally moves to its new building on the Victoria Embankment, and Abbott is able to extend the teaching of science throughout the school.

1884. *The Common Tradition of the Synoptic Gospels in the Text of the Revised Version* (with William G. Rushbrooke) is published. The object of the book, which is a translation of the first part of Rushbrooke's *Syntopicon*, is to present in English the "common tradition," the presumed common source used by the authors of Matthew, Mark, and Luke.

1884. *Flatland: A Romance of Many Dimensions* is published.

Edwin Abbott Abbott, c. 1884.

1885. *Francis Bacon: An Account of His Life and Works* is published.

1885. Abbott's American publisher, Roberts Brothers of Boston, issues *Flatland*. Roberts Brothers was one of the first American publishers to recognize the rights of foreign authors. Until the enactment of the International Copyright Act in 1891, other American publishers issued numerous unauthorized reprints of books by non-resident authors.

1886. Abbott presides at the annual meeting of the Teachers Guild (*The Times*, 22 March 1886, 6).

1886. *The Kernel and the Husk: Letters on Spiritual Christianity* is published, a book written in the form of letters from one who worships a "non-miraculous Christ" (Abbott) to an imaginary young man. Like all of Abbott's popular religious writing, it is based on three hypotheses: The synoptic gospels contain a number of doubtful stories set around a common nucleus of an original tradition. A belief in biblical miracles is not necessary for a true Christian faith. The attainment of knowledge through error and truth through illusion is an essential element of the divine method of revelation. In the twenty-fourth letter, Abbott refers to *Flatland* and Hinton's "*A Romance of the Fourth dimension*" (*sic*).

1887. Abbott's younger brother Sydney Wells Abbott, a librarian at the British Museum, dies. Edwin Abbott Parry describes him as "the ideal uncle for small nephews to worship. . . Rules and regulations failed to captivate his mind or restrain his conduct" (Parry 1932, 53–54). Abbott was not related to the well-known classical scholar, Evelyn Abbott

(1843–1901), despite the assertion made in the eleventh edition of *Ency-clopaedia Britannica* that Evelyn Abbott was his brother.

1887. Abbott's name heads the alphabetical list of members of the Cambridge Senate urging "that such steps as may be deemed necessary be taken to provide for the admission of duly qualified women to the degrees of the University of Cambridge" (*The Times*, 2 November 1887, 4).

1888. An American educator visiting the City of London School describes Abbott as "a slender, nervous man, about five feet four inches in height, with a most interesting and intellectual face" (Monroe 1889, 227).

1889. In 1887, the Court of Common Council proposed that the City of London School establish a "Commercial School" as well as a "Classical School." Abbott fought unsuccessfully to retain the position of classics, and losing this fight contributed to his decision to tender his resignation on 13 March 1889. He agrees to serve until January 1890 to allow the Court sufficient time to find a replacement (Hinde 1995, 64–66).

1889. *The Latin Gate: A First Latin Translation Book* is published.

1889. The Court of Common Council grants Abbott an annual pension of £400 over the objection of a member of the Court who observes that he "had enjoyed a salary of £1,100 for the last quarter of a century" (*The Times*, 14 February 1890, 11).

1890. Abbott delivers an address, "Civic and moral training in the schools," to the National Union of Teachers (*The Times*, 10 April 1890, 9, 12).

1890. In May, Abbott preaches the Commemoration Sermon at St John's College, Cambridge, on the text, "The truth shall make you free." He advises: "Avoid as far as you can the distractions of modern life, practice concentration, and be sometimes alone, not only with God in prayer, but with Shakespeare, with Plato, and with Wordsworth, whereby the cultivation of our intellects may increase your faith" (St. John's *Eagle* 16, 1891, 301).

1890. At a City of London School reunion dinner in October, Abbott is presented with his portrait painted by Hubert von Herkomer.

1890. Abbott delivers a lecture entitled "Illusions" at Toynbee Hall on October 4th. The accounts of this lecture in *The Times* are followed by letters to the editor from Abbott and the biologist T. H. Huxley, whom Abbott had cited as an example of a pessimist. The exchange ends amicably with Huxley declaring that Abbott "is the last person with whom I should wish to quarrel" (*The Times*, 11 October 1890, 7).

1890. In December, Richard Garnett, the keeper of printed books at the British Museum, offers Abbott a collection of letters written by Abbott's father to the librarian and writer, Edward Edwards. Abbott thanks Garnett for the offer but declines to accept the letters, saying:

> I have of late made it a rule to destroy all letters from relations and personal friends, and indeed almost all letters of every kind. I did this about two or three years ago, partly because I was moving into a small house from a large one and was straitened for room; partly because I have quite a genius for forgetting where papers are placed, and also for misplacing them, and partly from a dislike that they should fall into inappreciative hands (Abbott 1890).

1891. In May, Abbott writes two letters in support of Philip H. Calderon's recent painting, *St. Elizabeth of Hungary's Great Act of Renunciation*, which had been sharply criticized because it portrayed the thirteenth-century saint naked in a church in the presence of men (*The Times*, 21 May & 25 May 1891).

1891. *Philomythus: An Antidote against Credulity* is published. This severely critical examination of Cardinal Newman's famous essay on ecclesiastical miracles provoked outrage among the admirers of Newman, including R. H. Hutton, the editor of *The Spectator*, who reviewed it very unfavorably. Abbott replied with a long letter to *The Spectator*, which Hutton published with an editorial reply (25 April 1891, 590–593). Abbott wrote again but Hutton refused to publish the letter and declared the correspondence closed. Abbott then published at his own expense a seventy-five-page pamphlet titled *Newmanianism*, in which he included the rejected letter and recounted the entire episode. This pamphlet became the preface to the second edition of *Philomythus*.

1892. *The Anglican Career of Cardinal Newman* (2 vols.) is published. Undeterred by controversy, Abbott followed *Philomythus* with two bulky volumes in which he scrutinizes Newman's life and thought up to 1845, when he left the Anglican Church to become a Roman Catholic.

1892. Mary Abbott is placed in the first class of Part 1 of the Classical Tripos.

1893. *Dux Latinus: A First Latin Construing Book* is published.

1897. *The Spirit on the Waters: The Evolution of the Divine from the Human* is published. On the title page, Abbott publicly acknowledges for the first time his authorship of four books that were published anonymously: *Philochristus, Onesimus, The Kernel and the Husk*, and *Flatland*. In a section on the conception of God, Abbott describes the climactic visitation scene from *Flatland* and discusses the possible responses of a Flatlander, the most immediate of which might be to worship this being because of its "God-like" powers (Abbott 1897, 29–31).

1897. On May 21, the members of the University of Cambridge defeat a resolution that urged the conferring of the Bachelor of Arts degree upon qualified women. Abbott is among 661 voting *placet* (in favor of the resolution), his son is among the 1,707 voting *non-placet* (*Special Supplement to the Cambridge Review*, 3 June 1897, ii–iii).

1898. *St. Thomas of Canterbury: His Death and Miracles* is published. Abbott says that he was led to write this book because his study of the early narratives of Becket's death and miracles had revealed "parallelisms to problems of New Testament criticism."

1900. *Clue: A Guide through Greek to Hebrew Scripture* is published. The first volume of Abbott's "Diatessarica" series, an exhaustive study of the four Gospels.

1900. Abbott helps May Sinclair get her translation of Homer's "A hymn of Apollo" published in *Literature*. Twenty years later, she had become one of the most widely known woman authors in England.

1901. *Corrections of Mark Adopted by Matthew and Luke* is published.

1903. *From Letter to Spirit; an Attempt to Reach through Varying Voices the Abiding Word* is published.

1903. *Contrast, or a Prophet and a Forger* is published.

1904. *Paradosis, or in the Night in which He Was (?) Betrayed* is published.

1905. *Johannine Vocabulary: A Comparison of the Words of the Fourth Gospel with those of the Three* is published.

1906. In his retirement, Abbott avoided "the distractions of modern life"; indeed, after 1900 he rarely left Hampstead. In 1906, he wrote to Llewelyn Davies, "I have not been out of Hampstead for (I think) five years, and during these years I have worked three hundred and sixty-three days in each year, or thereabouts, and have been wonderfully well, memory (for words) never better" (Abbott 1906).

1906. *Johannine Grammar* is published. Whatever their opinions of Abbott's theological views, reviewers were unanimous in praising his *Johannine Vocabulary* and *Johannine Grammar* as profoundly important contributions to the study of the Gospel and Epistles of John. "For the professed scholar, they are indispensable. . . They will be found to be, what so many more popular books are not, a real and important addition to knowledge" (Sanday 1906, 155).

1906. *Silanus the Christian* is published. The purported autobiography of a young Roman disciple of the stoic philosopher Epictetus who is won over to Christianity. Unlike *Philochristus*, *Onesimus*, and *Flatland*, it is not written in an archaic style.

1907. *Apologia: An Explanation and Defence* is published, a collection of notes on the text of *Silanus the Christian*. In the Preface, Abbott identifies the "H.C." to whom *Flatland* is dedicated, and in the text he refers to *Flatland* twice (see Notes 22.23 and E.128).

1907. *Indices to Diatessarica: With a Specimen of Research* is published.

1907. *Notes on New Testament Criticism* is published. The reviewer of this book compares Abbott to Origen, the most important theologian and biblical scholar of the early Greek church. "There is the same extraordinary fertility of ideas, the same scholarly accuracy in their presentation, and the same difficulty for the rest of the world in deciding what can be assimilated and what cannot" (Sanday 1908, 114).

1907. *Revelation by Visions and Voices* is published (see Note 22.23).

1909. Herbert H. Asquith, the most famous of Abbott's students, becomes Prime Minister of the United Kingdom. In his autobiography, Asquith said, "Abbott had a natural gift for teaching, and in the everyday tasks of translation, it would be difficult to imagine a better equipped or more stimulating master. . . Above all, he had the sovereign gift of a great personality, at once austere and sympathetic, impressive and inspiring" (Asquith 1928, 11–12). Abbott reminisces on Asquith's career at the City of London School in Alderson (1905, 9–10).

1909. *The Message of the Son of Man* is published.

1910. *The Son of Man, or Contributions to the Study of the Thoughts of Jesus* is published. The previous six volumes in the "Diatessarica" series were printed in Cambridge at the University Press, but were published by Adam and Charles Black. This one and the seven that followed were published by Cambridge University Press.

1912. *Light on the Gospel from an Ancient Poet* is published. Abbott is made an honorary fellow of St. John's.

1913. *Miscellanea Evangelica, Vol. I,* and *The Fourfold Gospel, Introduction* are published. Abbott is made a Fellow of the British Academy.

1914. *The Fourfold Gospel, the Beginning* is published.

1915. *The Fourfold Gospel, the Proclamation of the New Kingdom* and *Miscellanea Evangelica, Vol. II (Christ's Miracles of Feeding)* are published.

1916. *The Fourfold Gospel, the Law of the New Kingdom* is published. "This new work helps the reader to a deepened appreciation of the learning and devotion of perhaps the most industrious English scholar of our time" (Knight 1916).

1917. *The Fourfold Gospel, the Founding of the New Kingdom* is published, the fourteenth and last volume of the "Diatessarica series," which comprises more than 7,100 pages.

Edwin Abbott Abbott, c. 1914.

1918. On his eightieth birthday, Abbott is presented with a tribute signed by several hundred men and women of distinction (*The Times*, 21 December 1918, 3).

1918. Abbott's last published paper, "Righteousness in the Gospels," appears in the *Proceedings of the British Academy* 8 (1917/1918), 351–364.

1919. Abbott's wife dies in February. He is bedridden from this time until his own death.

1926. *Flatland* is reprinted in England for the first time since 1885 by Sir Basil Blackwell, who bought the rights from Abbott's daughter.

1926. Abbott dies of influenza at his home, Wellside, Well Walk, Hampstead, on October 12. He is buried in Hampstead Cemetery, Fortune Green, on October 15; a former student, James G. Simpson, gives a moving funeral oration (Simpson 1926). Obituaries are found in (Obituary 1926a; 1926b).

Appendix B2. A "mathematical biography" of Edwin Abbott Abbott[4]

Edwin Abbott's father was the headmaster of the Marylebone Philological School and the author of several schoolbooks, most notably a *Handbook of Arithmetic and First Steps in Algebra*. There is no record of Abbott's early education, but it is likely that he was educated at home by his parents until he entered the City of London School when he was not quite twelve years old. There he came under the profound influence of the headmaster George F. W. Mortimer, who had earned first class honours in the classics at Queen's College, Oxford.

In the middle of the nineteenth century, Eton, Harrow, and Rugby devoted only three hours per week to mathematics, and this relative neglect of mathematics was common at other public schools. By contrast at CLS, which Abbott attended from 1850 to 1857, boys in the middle and upper classes (ages 14 to 18) spent about one-third of their class time on mathematics. Before the Taunton Commission in May 1865, Mortimer testified, "we have a very considerably higher mathematical standard, I suppose, than any other school in the kingdom." The curriculum of the sixth form mathematics class at CLS (twenty-three boys, average age $17^{1}/_{2}$) corroborates his testimony. During the 1864–1865 school year, this class studied Euclid's *Elements*, algebra, plane and spherical trigonometry, conic sections, Newton's *Principia* (Sections I–III), statics, dynamics, elementary hydrostatics, differential and integral calculus, optics, astronomy, and the theory of equations (Schools Inquiry Commission 1868a, 373; 1868b, 431).

Abbott, who took mathematics as far as differential equations at CLS, believed that administering such a heavy dose of mathematics to all students was a serious mistake. In "Confessions of a rationalist Christian," he comments on his experience as a student at CLS:

> Mathematics and not classics was our strong point, and under the pressure of our eccentric mathematical master, who must have been partially mad from the beginning and whom the treatment by his pupils probably made still madder, we covered more ground in mathematics than was ever attempted (as far as I know) in any English school before or since. All alike were forced through the mill of higher mathematics, with the result that several of our pupils took high mathematical degrees at Cambridge, while the majority had their education injured or spoiled (Douglas-Smith 1965, 125).

The eccentric master to whom Abbott refers was Robert Pitt Edkins, who taught at CLS from its founding in 1837 until his sudden death in November 1854. Edkins obtained only second class honours at Trinity

---

[4] A revised version of Lindgren and Richards (2009), which was based largely on Richards (1984).

College, Cambridge, in 1830. Eighteen years later, he had not published a single line on any subject; nonetheless, he was named the Gresham Professor of Geometry. In spite of his modest credentials and an utter inability to maintain classroom discipline, Edkins succeeded in producing a fair number of students who subsequently gained mathematical honours at the universities (Gresham professors 1848).

Abbott's mathematical master from 1854–1855 was Henry William Watson, who graduated from Trinity College, Cambridge, with first class honours in mathematics. He happened to be available at mid-year because he was in London studying law. Watson subsequently had a distinguished career as a mathematician and clergyman. He was succeeded by Francis Cuthbertson, a graduate of Corpus Christi College, Cambridge. Cuthbertson was a student at CLS with Abbott (1850–1851), taught him as a pupil (1856–1857), and served as Second Master during Abbott's entire tenure as headmaster (1856–1889), dying suddenly in December 1889, just before Abbott's retirement. He proved to be far more reasonable than Edkins, and just as effective in producing students who subsequently obtained first class honours in mathematics at Cambridge and Oxford (Gurney 1890).

Abbott treasured his days as CLS not only for the education and guidance he received from Mortimer and the other masters but perhaps even more for the enduring friendships he formed there. Two of his closest friends became mathematicians. Harold Candler was Abbott's dearest friend for more than sixty years. After graduating from Trinity College, Cambridge, he was the longtime mathematical master at Uppingham School under the famous headmaster Edward Thring. It is all but certain that Abbott consulted Candler when writing *Flatland*, which is dedicated to "the inhabitants of space in general and H.C. in particular." William S. Aldis, the son of a Baptist minister, was another of Abbott's close friends. Most of the students at CLS did not go on to one of the universities but left for business at age sixteen, and that was what Aldis's father had intended for him until Mortimer interceded. Instead, with the help of scholarships, he followed Candler to Trinity, whereas Abbott went to St. John's College, Cambridge.

When Abbott and Aldis entered Cambridge in 1857, they had two options for a B.A. degree: an Ordinary or an Honours degree.[5] The examination for the Honours degree was originally a mathematics examination called the Senate House (or Tripos) examination. A Classical Tripos was instituted in 1824, and Tripos examinations in other areas soon followed. By the mid-nineteenth century, the Tripos examinations and particularly the Mathematical Tripos had become extremely difficult, and competition among students was fierce.[6] Any undergraduate who wished to excel was

---

[5] At this time, approximately 40% took Honours degrees, 35% Ordinary degrees, and 25% no degree (Tullberg 1998, 194).

[6] Leonard Roth called the Mathematical Tripos "far and away the most difficult mathematical test that the world has ever known" (Roth 1971, 228).

compelled to study under the direction of a private coach for his entire undergraduate career. In his classic study of academic life at Victorian Cambridge, Sheldon Rothblatt points out that these coaches were not simply providing supplementary or remedial instruction; "they were the most important teachers in the university" (Rothblatt 1968, 198).

Part I of the Mathematical Tripos, which Abbott and Aldis took in January 1861, lasted three days. It included problems on Euclid and conics, algebra and trigonometry, statics and dynamics, hydrostatics and optics, Newton's *Principia*, and astronomy. On this part, the emphasis was on the methods of classical geometry used from the time of Archimedes up to the time of Isaac Newton, and the use of the calculus was forbidden (*Cambridge University Calendar* 1861, 11–12). It is noteworthy that the mathematics curriculum at the City of London School included every one of these topics.

After Part I, there was a ten-day break during which the examiners determined those candidates who had earned "mathematical honours." Only those so designated were admitted to the second part, which lasted five days. The problems on Part II, described as "pure mathematics and natural philosophy," were in fact predominantly problems of mathematical physics. At the conclusion, the examiners ranked the students according to their scores on the papers from all eight days and determined which of them had earned first, second, and third class honours. Those who earned first class honours were called "wranglers"; the highest-scoring wrangler was the senior wrangler, the next highest the second wrangler, and so on. Those earning second class honours were called the "senior optimes," and those earning third class honours the "junior optimes."

The honours list of 1861 included thirty-four wranglers, thirty-three senior optimes, and twenty-six junior optimes. Abbott was the seventh senior optime, and William Aldis the senior wrangler.[7] On Tripos examinations between 1857 and 1866, CLS graduates included two senior wranglers, two second wranglers, two third wranglers, a fourth wrangler, two sixth wranglers, and six other wranglers. Howard Candler was sixteenth wrangler in 1860. Of Abbott's three mathematics masters, Edkins was the twenty-fifth senior optime in 1830, Watson the second wrangler in 1850, and Cuthbertson the fourth wrangler in 1855.

Abbott was not required to take the Mathematical Tripos, but he took it because a second class in mathematics was a necessary condition of competing for a Chancellor's Medal for classics. Given his mathematical preparation at CLS, Abbott had every reason to believe that he could gain second class honours in mathematics without unduly interfering with his preparation for the Classical Tripos. We have no evidence that he studied

---

[7] Aldis was tutored by the most successful of all Cambridge coaches, Edward J. Routh, who coached 26 senior wranglers and nearly half of the 990 wranglers between 1862 and 1888 (Warwick 2003, 233).

William S. Aldis and Abbott c. 1861.

with a private coach for the Mathematical Tripos; his college tutor was Stephen Parkinson, who coached four senior wranglers.

After Cambridge, Abbott taught briefly at King Edward's School, Birmingham, and Clifton College, and then returned to CLS as headmaster in 1865. During the twenty-five years that he was headmaster, many of Britain's mathematical masters began to argue that Euclid's *Elements* was poorly suited for use in the schools. They advocated that new texts be written that would be clear, accessible to their pupils, yet rigorous.

Because of the British reliance on examinations as the way to judge academic merit, replacing the *Elements* with alternative textbooks involved not only writing and using such textbooks but also changing the types of questions posed on mathematical examinations, particularly those at Oxford and Cambridge. The success of such an enterprise necessitated the cooperation of Britain's secondary schoolteachers as well as the mathematical communities at the universities.

Despite these difficulties, the proposal to replace the *Elements* with alternative texts was considered seriously during the last third of the century. J. J. Sylvester's "Plea for the mathematician," delivered at the meeting of the British Association in 1869, included a fervent call to abandon the *Elements* as a textbook. In 1870, the Headmasters' Conference, of which Abbott was an active member, overwhelmingly passed a resolution calling for a general reevaluation of the *Elements*. Widespread concern culminated in the formation of the Association for the Improvement of Geometrical

Teaching, which held its first meeting at University College, London, in January 1871. Howard Candler was a member from the outset. Abbott and Cuthbertson attended the second meeting in 1872 and became members in 1873 (AIGT 1872, 9). Abbott remained a member until 1884.

Neither Cuthbertson nor Abbott advocated any major reform in the teaching of geometry. Cuthbertson was among those who wrote an alternative text, but his *Euclidian Geometry* (1874) was a conservative revision of the *Elements*.[8] Abbott observes that many excellent teachers "object to Euclid as being cumbrous, circuitous, and artificial." But he does not say that he is among those who object, and is content to leave "to specialists the task of suggesting better methods or textbooks" (Abbott 1883, 189). Before a parliamentary committee, Abbott testified that CLS had not experienced the same difficulties in teaching Euclid that his friend James M. Wilson had complained of at Rugby (Select Committee on Scientific Instruction 1868, 187).

The institutional development of the AIGT and its ultimate failure are chronicled in Brock (1975). For a complete account of the struggle to replace Euclid's *Elements* in English schools, see chapter 4 of Richards (1987).

Appendix B3. Abbott and the education of girls

For much of the nineteenth century, the majority of English girls had very little formal education. Those who were educated at all were taught by governesses or at small private schools; such schooling was considered a way for girls to obtain social skills rather than intellectual development, which was thought to diminish femininity (Burstyn 1980, 22).

In 1871, Frances Buss, headmistress of the highly successful North London Collegiate School (the first public day school for girls), established a second school, the Camden School for Girls, with lower fees to accommodate girls from less affluent families (Scrimgeour 1950; Burchell 1971). In that same year at a prize-day ceremony at the Camden School with the Mayor in the chair, Abbott "spoke very strongly on the duty of the Mayor and Corporation to provide for girls schools similar to those for their brothers" (Ridley 1895, 108–109). In November 1871, Maria Grey issued an earnest request to the readers of *The Times* for contributions to provide classrooms and furniture for the Camden School. In January 1872, her appeal having raised less than £50 at the same time that £60,000 was quickly raised for a new boys' school in London, she expressed her fear that "people do not care about the education of girls, and do not think it worth giving money to."[9]

---

[8] Nonetheless, Lewis Carroll criticized Cuthbertson's departure from Euclid in Carroll (1879).

[9] M. G. Grey to the Editor of *The Times*, 14 November 1871 and 1 January 1872. Edward Ellsworth's *Liberators of the Female Mind* (1979) is the most complete account of the activities

To rally support and raise funds for the education of girls beyond the elementary level, Maria Grey and Emily Sherriff together with others formed the National Union for the Improvement of the Education of Women of All Classes (later called the Women's Education Union). The most substantial achievement of the Women's Education Union was the incorporation in 1872 of the Girls' Public Day School Company (GPDSC), which sold shares to finance the rental, purchase, or construction of secondary schools for girls, modeled on the North London Collegiate School. Beginning with a school at Chelsea in January 1873, the GPDSC opened thirty-eight schools in the first twenty-five years after its incorporation (Goodman 2005). Abbott was on the platform at the public meeting in the Albert Hall that launched the GPDSC; he served on the Company's council and on the local committee of the GPDSC school in St. John's Wood, which opened in 1876.[10]

The second major challenge facing the Women's Educational Union was meeting the overwhelming need for qualified secondary teachers of girls. In response to persistent appeals by Maria Grey, Abbott agreed to serve on a committee of the Women's Educational Union, which would make provision for the training of secondary teachers.[11] In 1877, the committee was incorporated as the Teacher's Training and Registration Society with the goals of providing instruction in the science of education, offering opportunities for supervised practice teaching, and certifying the results of the training by competent examiners. Abbott signed the articles of incorporation of the Society, served as its first chairman, and was a long-serving member of its Council (Lilley 1981, 11).

In May 1878, the Society opened the Training College for Higher Grade Teachers in Girls' Schools in Bishopgate (named the Maria Grey College in 1886), a non-residential training school for secondary teachers. At the same time that the Training College was being established, Abbott was urging the Vice Chancellors of both Oxford and Cambridge to arrange for their universities to examine prospective teachers and issue certificates attesting to their qualifications.[12] In addition to individuals like Abbott and Frances Buss, women's groups and the Headmasters' Conference also petitioned the universities to provide teacher training. At last, the University of Cambridge appointed a Teacher Training Syndicate, which established lectures on the theory, history, and practice of teaching and set up a system for examining and certifying prospective teachers in these subjects (Hirsch and McBeth 2004, 68). Between July 1878 and October 1879, Abbott

of Maria Sherriff Grey and her sister Emily Sherriff on behalf of women's education in Britain.

[10] Education for girls, *The Times*, 8 June 1872, 5 (Magnus 1923, 93). Abbott's daughter Mary attended the St. John's Wood school from 1881 to 1886.

[11] Abbott to Maria Grey, July 1876 and 10 January 1877.

[12] Abbott to the Vice Chancellor of Cambridge, 8 November 1877 and 1 June 1878.

wrote at least eighteen letters to Oscar Browning, the secretary of the Cambridge Syndicate, on matters relating to the Bishopgate Training School and the Syndicate.[13] The Syndicate administered its first examination in June 1880; of the twenty-seven candidates who passed that examination, eighteen were students of the Bishopgate Training College (*The Times*, 12 July 1880, 5).

Despite long lives, both Emily Sherriff and Maria Grey spent many years in ill health. Maria Grey was forced to give up her activities on behalf of the Training College shortly after its formation. In a letter from Rome, she wrote to Francis Buss, "Dr. Abbott deserves more thanks than I can express... Will you tell him... something of what we both feel about his generous gift of time and thought" (Ridley 1895, 278).

Appendix B4. Abbott obituary

> DR. EDWIN ABBOTT.
>   A Famous Head Master, Scholar, and Critic.[14]

We regret to announce the death of Dr. Edwin Abbott, at his Hampstead home, in his 88th year.

Edwin Abbott Abbott, born in London on December 20, 1838, was educated at the City of London School under Dr. Mortimer, whom he succeeded as head master in 1865 after a brilliant career at St. John's College, Cambridge, of which, as senior classic and Chancellor's medallist, he was elected Fellow in 1861, and a short apprenticeship at King Edward's School, Birmingham, and Clifton College. He held the head-mastership for twenty-five years with a success to which the distinction of having been "under Abbott" still bears eloquent testimony.

The City of London School, though based upon the old foundation of John Carpenter, is essentially an institution of the Victorian era, having been opened by the Corporation as a place of secondary education for boys in 1837. Its early days had been unfortunate, and to Abbott's own teacher, Dr. Mortimer, belongs the credit of having rescued it from what might have been an inglorious career. But it was undoubtedly Abbott who

---

[13] Browning's papers at the King's College Archive Centre, Cambridge, include twenty-three letters from Abbott to Browning.

[14] *The Manchester Guardian*, 14 October 1926, 13. A natural candidate for the author of this elegant tribute is Charles E. Montague, a writer for the *Manchester Guardian*, who was educated at the City of London School and at Balliol College, Oxford. Nevertheless, Montague says that he did not write it: "Alas, I cannot take credit for that excellent notice in the *Manchester Guardian* of that prince of men, E. A. Abbott. To have been taught by him is the most fortunate thing in my life" (Elton 1929, 11). We believe that the author is Robert S. Conway, a classical scholar and philologist. Conway attended CLS and Gonville and Caius College, Cambridge. In 1893, he became professor of Latin at University College, Cardiff, and in 1903 Hulme Professor of Latin in the University of Manchester, where he remained until his retirement in 1929.

gave it a position among the great day schools of England that bade fair to rival the reputation with which ancient traditions and many centuries of success had clothed Westminster or St. Paul's. His successive sixth forms have given a generation of able men to the service of Church and State, and he lived to see one of his pupils, Herbert Asquith, Prime Minister of England. It was, perhaps, only with a small minority even of the pupils who owe most to him that "the Doctor" came into such close touch as to reveal to them something of the tenderness and passion of his character. To most there was always an atmosphere austere and awesome about the small figure, with its big white tie, black hair, keen, grey eye, and profile like nothing so I much as the prow of an eight. There was something of "the pure severity of perfect light" in his outward manner that subdued the ordinary boy. Forty years ago the head master was expected to be more or less of a remote being, and this was perhaps emphasised in the conditions of a day school. But his clear-cut translations of Thucydides, his expositions of Tennyson's "In Memoriam" or Bacon's essays, and, above all, his marvellous treatment of the Epistles of St. Paul are a possession for ever to those whom he taught not only how to write clearly but how to think logically and honestly and to prize the priceless treasure of their mother tongue.

It is not, however, as a schoolmaster that Dr. Abbott would himself have wished to be remembered. The real, we had almost said the passionate, interest of his life was the problem of presenting Christianity to the modern mind in a light that would ensure the permanence of traditional beliefs under conditions which seemed to make the acceptance of miracle impossible. It would be a mistake to identify him with the "New Theology," a movement that arose a quarter of a century after the date when his own attitude was determined, and that was far too superficial to satisfy his comprehensive mind. Reaching maturity in the sixties, he felt the pressure of physical science, as men felt it when Darwin, Huxley, and Tyndall were revolutionising their habits of thought, and the limits of natural law seemed, it may be, more definite and arbitrary than they do to-day. While, as it seemed to him, the majority were inclined to choose between the religion of Newman and the agnosticism of Herbert Spencer, he himself repudiated the dilemma. Complete accord with his attitude is not necessary to a deep reverence for the nobility and patience of the man, who set himself the titanic task of bringing a scholarship that has been rarely rivalled in this generation, a powerful but chastened imagination, a critical ability that had already proved its fine quality on the texts of Bacon and Shakespeare, a mind whose supreme instinct was to build, but never with untested material, to bear upon a task, which would have appalled most men and been realised by few.

It is the mark of a great teacher that, open his writings where you will, you become aware of his message, while at the same time there is no monotony of undiversified repetition. This is profoundly true of Abbott. Go where you will to the intricacies of his "Diatessarica," to his monograph on Beckett, to the prefaces of his sermons, to his close study of the Anglican career of Cardinal Newman, to his reconstructions of the mind of primitive believers in "Philochristus," "Onesimus," or "Silanus the Christian," and while you are impressed with the width of learning and variety of interest which illuminate even the dullest pages, you marvel at the wonderful tenacity of a purpose beside which the advancement of learning becomes insignificant. This purpose was not, as the careless might assume, to destroy impossible beliefs, but to bring all men to love, reverence, and worship the Founder of the Christian religion.

It is many years since the world was aware of Edwin Abbott. His latest pupils are already well advanced in the work of life. Those who remember him as a preacher are middle-aged men. He voluntarily withdrew from positions which would have kept him in the public eye, because his heart was set, not on popular distinction, but on public service. To those who knew the man and who are capable of estimating the brilliant, and varied qualities required for a literary activity and a critical study so broad and deep as that which is represented by the long list of his writings he will always remain a splendid and inspiring example, not unworthy to rank with those teachers who in one of the windows of his old school are described as "not of an age but for all time."

Recommended Reading

Abbott biography

Douglas-Smith, Aubrey E. 1965. *The City of London School*. 2nd ed. Oxford: Basil Blackwell.

Farnell, Lewis R. 2004. Abbott, Edwin Abbott (1838–1926), rev. Rosemary Jann, in *Oxford Dictionary of National Biography*, Oxford University Press.

Higher-dimensional geometry

Banchoff, Thomas F. 1990a. *Beyond the Third Dimension: Geometry, Computer Graphics, and Higher Dimensions*. New York: Scientific American Library.

Cajori, Florian. 1926. Origins of fourth dimension concepts. *Amer. Math. Monthly* 33 (October), 399–401.

Henderson, Linda Dalrymple. 1983. *The Fourth Dimension and Non-Euclidean Geometry in Modern Art*. Princeton: Princeton University Press.

Hinton, Charles H. and R. v. B. Rucker. 1980. *Speculations on the Fourth Dimension: Selected Writings of Charles H. Hinton*. New York: Dover Publications.

Kaku, Michio. 1994. *Hyperspace: A Scientific Odyssey through Parallel Universes, Time Warps, and the Tenth Dimension*. New York: Oxford University Press.

Manning, Henry P., ed. 1910. *The Fourth Dimension Simply Explained*. Reprint, New York: Dover, 2005.

Manning, Henry P. 1914. *Geometry of Four Dimensions*. New York: Macmillan.

Pickover, Clifford A. 1999. *Surfing through Hyperspace: Understanding Higher Universes in Six Easy Lessons*. New York: Oxford University Press.

Richards, Joan L. 1988. *Mathematical Visions: The Pursuit of Geometry in Victorian England*. Boston: Academic Press.

Rucker, Rudy v. B. 1977. *Geometry, Relativity, and the Fourth Dimension*. New York: Dover Publications.

Rucker, Rudy v. B. 1984. *The Fourth Dimension: Toward a Geometry of Higher Reality*. Boston: Houghton Mifflin Co.

Valente, K. G. 2008. Who will explain the explanation? The ambivalent reception of higher dimensional space in the British spiritualist press, 1875–1900. *Victorian Periodicals Review* 41, 124–149.

Weeks, Jeffrey R. 2002. *The Shape of Space*. 2nd ed. New York: Marcel Dekker.

Flatland general

Banchoff, Thomas F. 1991. Introduction to *Flatland*. Princeton: Princeton University Press.

Benford, Gregory. 1995. Introduction to *Flatland*. Norwalk, CT: Easton Press.

Burger, Dionys. 1965. *Sphereland: A Fantasy about Curved Spaces and an Expanding Universe*, translated from the Dutch by Cornelie J. Rheinboldt. New York: T. Y. Crowell.

Dewdney, Alexander K. 1984a. Introduction to *Flatland*. New York: New American Library.

Dewdney, Alexander K. 1984b. *The Planiverse: Computer Contact with a Two-Dimensional World*. Reprint, New York: Copernicus Books, 2001.

Gardner, Martin. 1969. Flatlands in *The Unexpected Hanging and Other Mathematical Diversions*. New York: Simon & Schuster.

Gilbert, Elliot L. 1991. "Upward not northward": *Flatland* and the quest for the new. *English Literature in Transition* 34, 391–404.

Jann, Rosemary. 1985. Abbott's *Flatland*: Scientific imagination and 'natural Christianity.' *Victorian Studies* 28, 473–490.

Jann, Rosemary. 2006. *Introduction to* Flatland. New York: Oxford University Press.

Lightman, Alan. 1998. *Introduction to* Flatland. New York: Penguin Books.

Smith, Jonathan. 1994. *Fact and Feeling: Baconian Science and the Nineteenth-Century Literary Imagination*. Madison: University of Wisconsin Press.

Stewart, Ian. 2002. *The Annotated Flatland: A Romance in Many Dimensions*. Cambridge, MA: Perseus Publishing.

Suvin, Darko. 1983. *Victorian Science Fiction in the UK: The Discourses of Knowledge and Power*. Boston: G. K. Hall, 370–373.

Plato's Cave

Sinaiko, Herman L. 1965. *Love, Knowledge, and Discourse in Plato; Dialogue and Dialectic in Phaedrus, Republic, Parmenides*. Chicago: University of Chicago Press, 167–184.

Classical Greece

Hornblower, Simon and Antony Spawforth, eds. 1996. *Oxford Classical Dictionary*. New York: Oxford University Press.

Turner, Frank M. 1981. *The Greek Heritage in Victorian Britain*. New Haven: Yale University Press.

Victorian Britain

Mitchell, Sally, ed. 1988. *Victorian Britain: An Encyclopedia*. New York: Garland Publishing.

Tucker, Herbert F., ed. 1999. *A Companion to Victorian Literature and Culture*. Malden, MA: Blackwell.

Young, George M. 1936. *Victorian England; Portrait of an Age*. Reprint, London: Phoenix Press, 2002.

References

Abbott, Edwin A. 1859. Review of Richard G. White's *The Works of William Shakespeare. The North American Review* 88, 244–253.

———. 1868a. The church and the congregation in Walter L. Clay, ed., *Essays in Church Policy*. London: Macmillan and Co., 158–191.

———. 1868b. The teaching of English. *Macmillan's Magazine* 18 (May), 33–39.

———. 1870. *A Shakespearian Grammar. An Attempt to Illustrate Some of the Differences between Elizabethan and Modern English. For Use in the Schools.* 3rd ed. London: Macmillan and Co.

———. 1871. On the teaching of the English language. *The Educational Times* 24 (February), 243–249; (March), 271–277.

———. 1872. Abbott to Alfred Marshall, 25 May. Trinity Library. Sidgwick/Add. Ms. c. 104/41.

———. 1874. Abbott to Macmillan and Co., 6 January. British Library, Macmillan correspondence, ADD 55114, f.19.

———. 1875a. *Cambridge Sermons Preached before the University*. London: Macmillan and Co.

———. 1875b. Abbott to George Eliot, 18 April. Leeds University Library, BC MS 19c.

———. 1877a. *Through Nature to Christ: or, the Ascent of Worship through Illusion to the Truth*. London: Macmillan and Co.

———. 1877b. *Bacon and Essex: A Sketch of Bacon's Earlier Life*. London: Seeley, Jackson, and Halliday.

———. 1877c. Abbott to Macmillan and Co., 23 January. British Library, Macmillan correspondence Add 55114 f.32–33.

———. 1877d. The training of teachers. *The Spectator*, 24 February, 246–247.

———. 1877e. Abbott to Macmillan and Co., 24 April. British Library, Macmillan correspondence, ADD 55114, f.36.

———. 1878. *Philochristus: Memoirs of a Disciple of the Lord*. London: Macmillan and Co.

———. 1882a. *Onesimus: Memoirs of a Disciple of St. Paul*. London: Macmillan and Co.

———. 1882b. Abbott to Gladstone, 16 September. British Library, Add 56450 f.82.

———. 1883. *Hints on Home Teaching*. London: Seeley, Jackson, and Halliday.

———. 1886. *The Kernel and the Husk: Letters on Spiritual Christianity*. London: Macmillan and Co.

———. 1888. Latin through English. *Journal of Education* 10 (1 August), 381–386.

———. 1890. Abbott to Richard Garnett, 11 December. Harry Ransom Center, University of Texas at Austin.

———. 1897. *The Spirit on the Waters: The Evolution of the Divine from the Human*. London: Macmillan and Co.

———. 1906. Abbott to J. Llewelyn Davies, 7 May. *From a Victorian Post-Bag: Being Letters Addressed to the Rev. J. Llewelyn Davies*, edited by Charles L. Davies. London: P. Davies, 1926, 59–60.

———. 1907a. *Apologia: An Explanation and Defence*. London: A. and C. Black.

———. 1907b. Revelation by visions and voices. *Essays for the Times*, No. 15, London: Francis Griffiths.

———. 1916. *The Fourfold Gospel. Section IV, The Law of the New Kingdom*. Cambridge University Press.

———. 1917. *The Fourfold Gospel. Section V, The Founding of the New Kingdom, or, Life Reached through Death*. Cambridge University Press.

Abbott, Edwin A. and Arthur J. Butler. 1884. The metaphysics of Flatland. *The Athenaeum* No. 2980 (6 December), 733.

Abbott, Edwin A. and John R. Seeley. 1871. *English Lessons for English People*. Reprint, Boston: Roberts Brothers, 1872.

Alderson, J. P. 1905. *Mr. Asquith*. London: Methuen and Co.

Aldis, Amy L. 1940. A brief biography of William Steadman Aldis. Typescript in the Aldis Papers, University of Auckland Library.

Aristotle. 1984. *The Complete Works of Aristotle: The Revised Oxford Translation*. Edited by Jonathan Barnes. Princeton: Princeton University Press.

> *On the Heavens* translated by J. L. Stocks.
> *Politics* translated by Benjamin Jowett.
> *Posterior Analytics* translated by Jonathan Barnes.

Asquith, Herbert H. 1928. *Memories and Reflections, 1852–1927*, vol. 1. Boston: Little, Brown, and Company.

Association for the Improvement of Geometrical Teaching. 1872. Second annual report.

Bacon, Francis. 1860. *The Works of Francis Bacon*, Vol. XIII. Edited and translated by J. Spedding, R. L. Ellis, and D. D. Heath. Boston: Brown and Taggard.

———. 1965. *The Advancement of Learning*. Edited by G. W. Kitchin. London: Dent.

Bagehot, Walter. 1872. *Physics and Politics, or, Thoughts on the Application of the Principles of "Natural Selection" and "Inheritance" to Political Science*. Reprint, New York: D. Appleton, 1906.

Banchoff, Thomas F. 1990b. From *Flatland* to hypergraphics. *Interdisciplinary Science Reviews* 15, 364–372.

Barnett, Henrietta. 1919. *Canon Barnett, His Life, Work, and Friends*, Vol. II. Boston: Houghton Mifflin Co.

Booth, William. 1890. *In Darkest England, and the Way Out*. New York: Funk and Wagnalls.

Borg, James M. 2004. Abbott, Edwin (1808–1882) in *Oxford Dictionary of National Biography*. Oxford University Press.

Boyer, Carl B. 1949. *The History of the Calculus and its Conceptual Development*. New York: Dover Publications.

Brock, W. H. 1975. Geometry and the universities: Euclid and his modern rivals 1860–1901. *History of Education* 4, 21–25.

Burchell, D. 1971. *Miss Buss' Second School*. London: Frances Mary Buss Foundation.

Burkert, Walter. 1987. *Ancient Mystery Cults*. Cambridge, MA: Harvard University Press.

Burstyn, J. N. 1980. *Victorian Education and the Ideal of Womanhood*. London: Croom Helm.

Butler, Arthur J. 1884. Review of *Flatland*. *The Athenaeum* No. 2977 (November 15), 622.

Byrne, Oliver. 1847. *The First Six Books of the Elements of Euclid in which Coloured Diagrams and Symbols are Used Instead of Letters for the Greater Ease of Learners*. London: William Pickering.

Cambridge University Calendar for the Year 1861. 1861. Cambridge: Deighton, Bell, and Co.

Camden Medal. 1860. *The Ecclesiastical Gazette* (12 June), 306.

Campbell, Lewis. 1898. *Religion in Greek literature; a Sketch in Outline*. London: Longmans, Green, and Co.

Carroll, Lewis. 1874. *Dynamics of a Parti-cle*. Oxford: James Parker and Co.

——. 1879. *Euclid and his Modern Rivals*. London: Macmillan and Co.

City of London School. 1882. *The Nation* 35, 352.

Conway, Robert S. 1926. Dr Abbott as teacher, *The Manchester Guardian* (15 October), 20.

Cussans, John E. 1893. *Handbook of Heraldry*. London: Chatto and Windus.

Davidoff, Leonore. 1973. *The Best Circles; Society, Etiquette and the Season*. London: Croom Helm.

Denison, David. 1993. *English Historical Syntax: Verbal Constructions*. London: Longman.

Dewdney, Alexander K. 2002. Review of *The Annotated Flatland*. *Notices of the American Mathematical Society* 49 (November), 1262.

Disraeli, Benjamin. 1845. *Sybil; or The Two Nations*. Reprint, Oxford University Press, 1926.

Douglas, Roy. 1999. *Taxation in Britain since 1660*. New York: St. Martin's Press.

Duff, Michael J. 2001. The world in eleven dimensions: a tribute to Oskar Klein. Oskar Klein Professorship Inaugural Lecture, University of Michigan, 16 March.

Eddington, Arthur S. 1921. *Space, Time, and Gravitation: An Outline of the General Relativity Theory*. Cambridge University Press.

Elton, Oliver. 1929. *C.E. Montague, a Memoir*. London, Chatto & Windus.

Einstein, Albert. 1921. *Relativity: The Special and General Theory*. Translated by R. W. Lawson. New York: Henry Holt and Co.

Farnell, Lewis R. 1926. Dr. Abbott: A former student's tribute. *The Times*, 18 October, 10.

Fawcett, Millicent G. 1870. The electoral disabilities of women. *The Fortnightly Review* 13 (May), 622–632.

Fechner, Gustav T. [Dr. Mises, pseud.] 1875. *Kleine Schriften*. Leipzig: Breitkopf and Härtel.

Field, Judith V. 1997. *The Invention of Infinity: Mathematics and Art in the Renaissance*. Oxford University Press.

Fitch, J. G. 1876. The universities and the training of teachers. *The Contemporary Review* 29, 95–116.

Fowler, Henry W. and Francis G. Fowler. 1906. *The King's English*. Oxford: Clarendon Press, 198–200.

Fowler, Henry W. and Ernest Gowers. 1965. *A Dictionary of Modern English Usage*. New York: Oxford University Press.

Gagarin, Michael and David Cohen, eds. 2005. *The Cambridge Companion to Ancient Greek Law*. Cambridge University Press.

Galton, Francis. 1869. *Hereditary Genius: An Inquiry into Its Laws and Consequences*. London: Macmillan and Co.

———. 1875. The history of twins, as a criterion of the relative powers of nature and nurture. *Fraser's Magazine* 92, 566–576.

———. 1905. Studies in eugenics. *Amer. J. Sociology* 11 (July), 11–25.

Garnett, William. 1932. Letter to the editor of *The Times*, 15 September, 13.

Gibson, Gabriella and Ian Russell. 2006. Flying in tune: sexual recognition in mosquitoes. *Current Biology* 16 (11 July), 1311–1316.

Girton College Register, 1869–1946. 1948. Cambridge: private printing.

Granville, William A. 1922. *The Fourth Dimension and the Bible*. Boston: R. G. Badger.

Gladstone, W. E. 1882. Gladstone to Abbott, 17 September. *The Gladstone Diaries*, Vol. 10. Edited by H. C. G. Mathew. Oxford University Press, 1994, 336.

Goodman, Joyce F. 2005. Girls' Public Day School Company. *Oxford Dictionary of National Biography*, online edition, Oxford University Press.

Grattan-Guinness, Ivor. 1997. *The Norton History of the Mathematical Sciences*. New York: W. W. Norton and Co.

The Gresham professors. 1848. *The Mechanics Magazine, Museum, Register, Journal, and Gazette* 49, 114–118.

Gurney, Henry P. 1890. In memoriam, *The City of London School Magazine* 19 (March), 3–9.

Hamilton, Edith, trans. 1937. *Three Greek Plays: Prometheus Bound, Agamemnon, The Trojan Women*. New York: W. W. Norton and Company.

Helmholtz, Hermann von. 1876. The origin and meaning of geometrical axioms. *Mind* 1, 301–321.

Hinde, Thomas. 1995. *Carpenter's Children: The Story of the City of London School*. London: James and James.

Hinton, Charles. H. 1880. What is the fourth dimension? *The University Magazine* [Dublin] 96, 15–34. Reprinted in *The Cheltenham Ladies College Magazine* 8 (1883), 31–52.

———. 1886. *Scientific Romances: First Series*. London: Swan Sonnenschein.

———. 1888. *A New Era of Thought*. (Part II was corrected and supplemented by A. B. Stott and H. J. Falk.) London: Swan Sonnenschein and Co.

———. 1907. *An Episode in Flatland; or How a Plane Folk Discovered the Third Dimension; to which is Added an Outline of the History of Unaea*. London: Swan Sonnenschein.

Hirsch, P. and M. McBeth. 2004. *Teacher Training at Cambridge: The Initiatives of Oscar Browning and Elizabeth Hughes*. London: Woburn Press.

Honey John R. de S. 1977. *Tom Brown's Universe: The Development of the Victorian Public School*. London: Millington.

Hopkins, Jasper. 1985. *Nicholas of Cusa on Learned Ignorance*: A Translation and an Appraisal of *De Docta Ignorantia*. 2nd ed. Minneapolis: A. J. Benning.

Houstoun, Robert A. 1930. *A Treatise on Light*. New York: Longmans, Green, and Co.

Ifrah, Georges. 2000. *The Universal History of Numbers: From Prehistory to the Invention of the Computer*. Translated by David Bellos, E. F. Harding, Sophie Wood, and Ian Monk. New York: J. Wiley.

James, Henry. 1905. *English Hours*. Boston: Houghton, Mifflin, and Co.

Jammer, Max. 1969. *Concepts of Space; the History of Theories of Space in Physics*. 2nd ed. Cambridge, MA: Harvard University Press.

Janich, Peter. 1992. *Euclid's Heritage: Is Space Three-Dimensional?* Dordrecht: Kluwer Academic Publishers.

Jones, Henry F. 1968. *Samuel Butler: Author of Erewhon (1835–1902), A Memoir*. New York: Octagon Books.

Jowett, Benjamin. 1879. Letter to Abbott, 2 June. *Letters of Benjamin Jowett, M.A.*, Vol. 3, edited by Evelyn Abbott and Lewis Campbell. New York: E. P. Dutton, 1899, 206.

Kearney, Richard. 1988. *The Wake of Imagination: Toward a Postmodern Culture*. Minneapolis: University of Minnesota Press.

Kincses, János. 2003. The determination of a convex set from its angle function. *Discrete Computational Geometry* 30 (2), 287–297.

Klein, Jacob. 1989. *A Commentary on Plato's Meno*. Chicago: University of Chicago Press.

Knight, Samuel K. 1916. Review of *The Fourfold Gospel. Section IV*. *The Times Literary Supplement* (13 April), 173.

Lake, Paul. 2001. The shape of poetry in Kurt Brown, ed. *The Measured Word*: On Poetry and Science. Athens: University of Georgia Press.

Lamarck, Jean-Baptiste. 1984. *Zoological Philosophy: An Exposition with Regard to the Natural History of Animals*. Translated by Hugh Elliot. University of Chicago Press.

Levi, Leone. 1885. *Wages and Earnings of the Working Classes*. London: J. Murray.

Lilley, I. M. 1981. *Maria Grey College, 1878–1976*. Twickenham: West London Institute of Higher Education.

Lindberg, David C. 1976. *Theories of Vision from Al-Kindi to Kepler*. Chicago: University of Chicago Press.

Lindgren, William F. and Joan L. Richards. 2009. Edwin Abbott and the mathematics of *Flatland*. *Notices of the American Mathematical Society* 56 (January), 185.

Lloyd, Genevieve. 1984. *The Man of Reason: "Male" and "Female" in Western Philosophy*. Minneapolis: University of Minnesota Press.

Lovejoy, Arthur O. 1960. *The Great Chain of Being*: A Study of the History of an Idea. New York: Harper and Row.

MacKinnon, Flora. I., ed. 1925. *Philosophical Writings of Henry More*. Oxford University Press.

Magnus, L. 1923. *The Jubilee Book of the Girls' Public Day School Trust*. Cambridge University Press.

Manners and Tone of Good Society by a Member of the Aristocracy. Or, Solecisms to Be Avoided. 1879. London: F. Warne and Co.

Mathews, Nieves. 1996. *Francis Bacon: The History of a Character Assassination.* New Haven: Yale University Press.

McMullin, Ernan. 2003. Van Fraassen's unappreciated realism. *Philosophy of Science* 70 (July), 455–478.

Menger, Karl. 1943. What is dimension? *Amer. Math. Monthly* 50 (January), 2–7.

Michael, Ian. 1987. *The Teaching of English: From the Sixteenth Century to 1870.* Cambridge University Press.

Miscellaneous. 1876. *Transactions of the National Association for the Promotion of Social Science.* London: Longmans, Green and Co.

Monroe, H. E. 1889. Visiting English schools. *Education* 10 (December), 227–229.

Morley, Tom. 1985. A simple proof that the world is three-dimensional. *SIAM Review* 27 (March), 69–71.

Nabokov, Vladimir V. 1980. *Lectures on Literature.* New York: Harcourt Brace Jovanovich.

Netz, Reviel. 1999. *The Shaping of Deduction in Greek Mathematics: A Study in Cognitive History.* Cambridge University Press.

A new philosophy. 1877. *City of London School Magazine* 1 (December), 277–283.

Obituary of E. A. Abbott. 1926a. *The Times*, 13 October, 19.

Obituary of E. A. Abbott. 1926b. *Manchester Guardian*, 14 October, 13.

Obituary of Edwin Abbott. 1952. *Jesus College Cambridge Society, Forty-Eighth Annual Report.* Cambridge University Press, 16–17.

Oresme, Nicole and Marshall Clagett, ed. 1968. *Nicole Oresme and the Medieval Geometry of Qualities and Motions; a Treatise on the Uniformity and Difformity of Intensities known as Tractatus de configurationibus qualitatum et motuum.* Madison: University of Wisconsin Press.

Ouspensky, P. D. 1997. *A New Model of the Universe.* Mineola, NY: Dover Publications.

Paley, William. 1802. *Natural Theology: Or, Evidences of the Existence and Attributes of the Deity.* Reprinted ed. by M. D. Eddy and D. Knight. Oxford University Press, 2006.

Parker, Robert B. 1983. *Miasma: Pollution and Purification in Early Greek Religion.* Oxford: Clarendon Press.

Parry, Edwin A. 1932. *My Own Way: An Autobiography.* London: Cassell and Co.

Philip, J. A. 1966. The "Pythagorean theory" of the derivation of magnitudes. *Phoenix* 20 (1), 32–50.

Plato. 1963. *The Republic.* 2 vols. Translated by Paul Shorey. Cambridge, MA: Harvard University Press.

——. 1999. *Symposium* in *Great Dialogues of Plato.* Translated by W. H. D. Rouse, E. H. Warmington, and P. G. Rouse. New York: Signet Classic.

——. 2000. *Timaeus.* Translated by Donald J. Zeyl. Indianapolis, IN: Hackett Publishing Company.

Plescia, Joseph. 1970. *The Oath and Perjury in Ancient Greece.* Tallahassee: Florida State University Press.

Pomeroy, Sarah B. 1994. *Xenophon, Oeconomicus: A Social and Historical Commentary.* Oxford: Clarendon Press.

Proctor, Richard A. 1884. Dream-space. *The Gentleman's Magazine* 256 (January), 35–46.

———. 1887. Gossip. *Knowledge*, (1 March), 116.

Reichenbach, Hans. 1958. *The Philosophy of Space and Time*. Translated by Maria Reichenbach and John Freund. New York: Dover Publications.

Report from the Schools Inquiry Commission. 1868a. *Parliamentary Papers*, 1867–1868, Vol. 28 (Reports of the commissioners, Vol. 4).

Report from the Schools Inquiry Commission. 1868b. *Parliamentary Papers*, 1867–1868, Vol. 28 (General reports by assistant commissioners, Vol. 7).

Report from the Select Committee on Scientific Instruction. 1868. *Parliamentary Papers*, 1867–1868, Vol. 15 (Reports from committees, Vol. 10).

Review of *Shakespearian Grammar*. 1870. *The Spectator* 43, 1551.

Richards, Joan L. 1984. Edwin Abbott Abbott's *Flatland* and the British mathematical community. Paper delivered at Centenary Conference on *Flatland*. Brown University, 13 October.

Richards, Robert J. 1987. *Darwin and the Emergence of Evolutionary Theories of Mind and Behavior*. Chicago: University of Chicago Press.

Ridley, A. E. 1895. *Frances Mary Buss and Her Work for Education*. London: Longmans, Green and Co.

Ridley, Matt. 2003. *Nature Via Nurture: Genes, Experience, and What Makes Us Human*. New York: HarperCollins.

Roberts, Samuel. 1882. Remarks on mathematical terminology, and the philosophic bearing of recent mathematical speculations concerning the realities of space. *Proceedings of the London Mathematical Society* 14 (November), 12.

Rodwell, George F. 1873. On space of four dimensions. *Nature* 8 (1 May), 8–9.

Romanes, George J. 1887. Mental Differences between Men and Women. *Nineteenth Century* 21 (May), 654–672.

Roth, Leonard. 1971. Old Cambridge days. *Amer. Math. Monthly* 78, 223–236.

Rothblatt, Sheldon. 1968. *The Revolution of the Dons, Cambridge, and Society in Victorian England*. New York: Basic Books.

Sanday, William. 1906. Review of Johannine Vocabulary and Johannine Grammar, *The Times Literary Supplement* (4 May), 154–155.

———. 1908. A modern Origen. *The Times Literary Supplement* (9 April), 114–115.

———. 1917. A great work completed. *Times Literary Supplement* (27 September), 460.

Schlatter, Mark D. 2006. How to view a Flatland painting. *The College Mathematics Journal* 37 (March), 114–120.

Scrimgeour, R. M., ed. 1950. *The North London Collegiate School, 1850–1950: Essays in Honour of the Centenary of the Frances Mary Buss Foundation*. Oxford University Press.

Schwab, Ivan R. 2004. Flatlanders. *British J. Ophthalmology* 88 (August), 988.

Simpson, J. G. 1926. Funeral address. *The City of London School Magazine* 49, 119–121.

Smith, William, ed. 1878. *A Dictionary of Greek and Roman Antiquities*. London: John Murray.

Solomon, Alan D. 1992. Pick a number: What Edwin Abbott did not know about Flatland. *Oak Ridge National Laboratory Review* 25 (2).

Spencer, Herbert. 1851. *Social Statics: Or, the Conditions Essential to Human Happiness Specified, and the First of Them Developed*. Reprint, New York: D. Appleton and Co, 1865.

———. 1873. *The Study of Sociology*. Reprint, New York: D. Appleton and Co, 1874.

Steggall, J. E. A. 1926. In memoriam. *The City of London School Magazine* 49, 121–124.

Stillwell, John. 2001. The story of the 120-cell. *Notices of Amer. Math. Soc.* 48 (January), 17–24.

The Student's Guide to the University of Cambridge. 1866. 2nd ed. Cambridge: Deighton, Bell.

Suvin, Darko. 1979. *Metamorphoses of Science Fiction: On the Poetics and History of a Literary Genre*. New Haven: Yale University Press.

Sylvester, James J. 1869. A plea for the mathematician. *Nature* 1 (30 December), 237–239.

———. 1884. Sylvester to Arthur Cayley, 2 November 1884. Reproduced in Karen H. Parshall, *J. J. Sylvester: Life and Work in Letters*, (1998), 253–255.

Tucker, Robert. 1884. Review of *Flatland*. *Nature* 31 (27 November), 76–77.

Tullberg, Rita McWilliams. 1998. *Women at Cambridge*. Cambridge University Press.

Turner, Frank M. 1974. *Between Science and Religion: The Reaction to Scientific Naturalism in Late Victorian England*. New Haven: Yale University Press.

Universities Commission Report. 1874. *Nature* 10 (15 October; 22 October), 475–476; 495–496.

Valente, K. G. 2004. Transgression and transcendence: *Flatland* as a response to 'A new philosophy.' *Nineteenth-Century Contexts* 26, 61–77.

Verne, Jules, Walter J. Miller, and Frederick P. Walter. 1993. *Jules Verne's Twenty Thousand Leagues under the Sea: The Definitive Unabridged Edition Based on the Original French Texts*. Annapolis, MD: Naval Institute Press.

Walsby, Anthony E. 1980. A square bacterium. *Nature* 283 (January), 69–71.

Waltershausen, Wolfgang S. v. 1856. *Gauss zum Gedächtnis* (*In Memory of Gauss*). Leipzig: S. Hirzel.

Ward, Mary Arnold. 1889. An appeal against female suffrage. *The Nineteenth Century* 25 (June), 781–788.

Warwick, Andrew. 2003. *Masters of Theory: Cambridge and the Rise of Mathematical Physics*. Chicago: University of Chicago Press.

Webb, Beatrice P. 1926. *My Apprenticeship*. New York: Longmans, Green and Co.

Wells, H. G. 1937. Wells to J. B. Priestley, 27 February. Harry Ransom Center, University of Texas at Austin.

———. 1952. The Plattner story in *28 Science Fiction Stories of H. G. Wells*. New York: Dover Publications, 441–461.

White, William H. 1915. *Last Pages from a Journal*. Edited by D. V. White. Oxford University Press.

Williams, David. 1850. *Composition, Literary and Rhetorical, Simplified*. London: W and T. Piper.

Woolf, Virginia. 1967. *Collected Essays*, Vol. 2. New York: Harcourt, Brace and World.

Index of Defined Words

Index